Fluoride and the Oral Environment

Monographs in Oral Science

Vol. 22

Series Editors

A. Lussi Bern
M.C.D.N.J.M. Huysmans Nijmegen
H.-P. Weber Boston, Mass.

Fluoride and the Oral Environment

Volume Editor

Marília Afonso Rabelo Buzalaf Bauru

50 figures, 31 in color, and 24 tables, 2011

Basel · Freiburg · Paris · London · New York · New Delhi · Bangkok ·
Beijing · Tokyo · Kuala Lumpur · Singapore · Sydney

'This monograph is dedicated to Gary Milton Whitford, my master and friend, who guided me through the avenues of Fluoride Research.'
Marília Afonso Rabelo Buzalaf

Marília Afonso Rabelo Buzalaf
Department of Biological Sciences
Bauru Dental School, University of São Paulo
Al. Octávio Pinheiro Brisolla, 9-75
17012-901
Bauru-SP (Brazil)

This volume received generous financial support from Procter & Gamble

Library of Congress Cataloging-in-Publication Data

Fluoride and the oral environment / volume editor, Marília Afonso Rabelo
Buzalaf Bauru.
    p. ; cm. -- (Monographs in oral science, ISSN 0077-0892 ; v. 22)
  Includes bibliographical references and indexes.
  ISBN 978-3-8055-9658-9 (hard cover : alk. paper) -- ISBN 978-3-8055-9659-6
(e-ISBN)
  1. Dental caries--Prevention. 2. Fluorides--Therapeutic use. I. Buzalaf,
Marília Afonso Rabelo. II. Series: Monographs in oral science ; v. 22.
0077-0892
  [DNLM: 1. Fluorides, Topical--therapeutic use. 2. Dental
Caries--prevention & control. 3. Fluorosis, Dental. W1 MO568E v.22 2011 /
QV 282]
  RK331.F558 2011
  617.6'7--dc23
                                        2011013013

Bibliographic Indices. This publication is listed in bibliographic services, including MEDLINE/Pubmed.

© Copyright 2011 by S. Karger AG, P.O. Box, CH–4009 Basel (Switzerland)
www.karger.com
Printed in Switzerland on acid-free and non-aging paper (ISO 9706) by Reinhardt Druck, Basel
ISSN 0077–0892
ISBN 978–3–8055–9658–9
e-ISBN 978–3–8055–9659–6

Contents

List of Contributors

Marília Afonso Rabelo Buzalaf
Department of Biological Sciences
Bauru Dental School, University of São Paulo
Al. Octávio Pinheiro Brisolla, 9-75
17012-901 Bauru-SP (Brazil)
Tel. +55 14 3235 8346, E-Mail mbuzalaf@fob.usp.br

Pamela DenBesten
Department of Orofacial Sciences, School of Dentistry
University of California, San Francisco
513 Parnassus Avenue
San Francisco, CA 94143 (USA)
Tel. +1 415 502 7828
E-Mail Pamela.DenBesten@ucsf.edu

Heitor Marques Honório
Department of Pediatric Dentistry
Orthodontics and Public Health
Bauru Dental School, University of São Paulo
Al. Octávio Pinheiro Brisolla, 9-75
17012-101 Bauru-SP (Brazil)
Tel. +55 14 3235 8256, E-Mail heitorhonorio@usp.br

Steven Marc Levy
Departments of Preventive & Community Dentistry
and Epidemiology, University of Iowa
College of Dentistry
Iowa City, IA 52242-1010 (USA)
Tel. +1 319 335 7185, E-Mail steven-levy@uiowa.edu

Wu Li
Department of Orofacial Sciences, School of Dentistry
University of California, San Francisco
513 Parnassus Avenue
San Francisco, CA 94143 (USA)
Tel. +1 415 476 1037, E-Mail wu.li@ucsf.edu

Adrian Lussi
Department of Preventive, Restorative and
Pediatric Dentistry, University of Bern
Freiburgstrasse 7
CH-3010 Bern (Switzerland)
Tel. +41 31 632 25 10
E-Mail adrian.lussi@zmk.unibe.ch

Ana Carolina Magalhães
Department of Biological Sciences
Bauru Dental School, University of São Paulo
Al. Dr. Octávio Pinheiro Brisolla 9-75
17012-190 Bauru-SP (Brazil)
Tel. +55 14 3235 8247, E-Mail acm@usp.br

Juliano Pelim Pessan
Department of Pediatric Dentistry and Public Health
Araçatuba Dental School, São Paulo State University
Rua José Bonifácio, 1193
16015-050 Araçatuba - SP (Brazil)
Tel. +55 18 3636 3274, E-Mail jpessan@foa.unesp.br

Daniela Rios
Department of Pediatric Dentistry
Orthodontics and Public Health
Bauru Dental School, University of São Paulo
Al. Dr. Octávio Pinheiro Brisolla 9-75
17012-190 Bauru-SP (Brazil)
Tel. +55 14 32358218, E-Mail daniriosop@yahoo.com.br

Colin Robinson
Division of Oral Biology
Leeds Dental Institute, University of Leeds
Clarendon Way
LS2 9LU, Leeds (UK)
Tel. +44 113 343 6159, E-Mail c.robinson@leeds.ac.uk

Andrew John Rugg-Gunn
Morven
Boughmore Road
Sidmouth
Devon EX10 8SH (UK)
Tel. +44 1395 578746
E-Mail andrew@rugg-gunn.net

Fábio Correia Sampaio
Federal University of Paraiba
Health Science Centre
Department of Clinical and Social Dentistry
58051-900 João Pessoa (Brazil)
Tel. +55 83 3216 7795
E-Mail fabios@ccs.ufpb.br

Jacob Martien ten Cate
Department of Cariology
Endodontology and Pedodontology
Academic Center for Dentistry Amsterdam (ACTA)
Gustav Mahlerlaan 3004
NL-1081 LA Amsterdam (The Netherlands)
Tel. +31 20 518 8440, E-Mail J.t.Cate@acta.nl

Kyriacos Jack Toumba
Division of Child Dental Health
Leeds Dental Institute, University of Leeds
Clarendon Way
LS2 9LU Leeds (UK)
Tel. +44 113 343 6141, E-Mail k.j.toumba@leeds.ac.uk

Alberto Enrique Villa
Institute of Nutrition and Food Technology
INTA, University of Chile
Macul
Santiago 783–0480 (Chile)
Tel. +56 2 2212249
E-Mail avilla@inta.uchile.cl

Gerald Lee Vogel
American Dental Association Foundation
Paffenbarger Research Center
100 Bureau Drive Stop 8546
National Institute of Standards and Technology
Gaithersburg MD 20899-8546 (USA)
Tel. +1 301 975 6821, E-Mail jvogel@nist.gov

Gary Milton Whitford
Department of Oral Biology
School of Dentistry
Medical College of Georgia
1120 15th Street
Augusta, GA 30912 (USA)
Tel. +1 706 721 0388, E-Mail gwhitfor@mcg.edu

Annette Wiegand
Department of Preventive Dentistry
Periodontology and Cardiology
University of Zurich
Plattenstrasse 11
CH-8032 Zurich (Switzerland)
Tel. +41 44 6343412
E-Mail annette.wiegand@zzm.uzh.ch

Foreword

Working and publishing in the field of skeletal and dental tissues for the past 40 years, in particular on the biology of dental enamel, it became apparent to me at a very early stage that fluoride, a minor tissue constituent, was an inextricably important aspect of this area of study. Indeed the effects of fluoride seemed at odds with the extremely small amounts present. Also, unlike other important minor components of the skeletal and dental tissue mineral, such as carbonate and magnesium, fluoride concentrations vary widely and depend to a great extent on exposure to external sources.

Fluoride came to prominence by virtue of its effect on skeletal tissue development, particularly in relation to environmental exposure. In cases of exposure to relatively high concentrations, its presence during formation – as well as its direct incorporation into the skeletal mineral phase – led to pathological changes in both skeletal and dental tissues. Tooth enamel was found to be particularly sensitive in this respect. The effects can be profound since pathology related to high levels of fluoride exposure involves changes in both tissue structure as well as chemistry.

However, a paradox emerged from this field of study in which it became clear that exposure to smaller amounts of fluoride, while often leading to changes in the dental tissues, conferred considerable protection against the most widely spread and costly of diseases – dental caries. The protection was dramatic and was first ascribed to fluoride-induced changes to the tooth tissues during their development. This concept, however, was later challenged as topical exposure to fluoride in the oral environment was shown to be extremely effective in reducing dental caries. The role of developmentally acquired fluoride in this respect remains intriguingly open to question.

Such an important advantageous clinical effect, together with the concomitant possibility of pathological change, led to a wide range of intense investigations. These centered on how fluoride is obtained from the diet, how it is dealt with after absorption and also its interaction with the calcium phosphate/apatite phase of dental tissues.

The chemistry of biological calcium phosphates is, however, very complex. As a result of this, the deposition and behavior of the highly substituted and defect calcium hydroxyapatite crystals of the skeletal and dental tissues has received an enormous amount of attention. The interaction of fluoride with this system added further to this complexity, and as a consequence studies of fluoride and skeletal and dental mineral have generated a vast literature.

With the obvious potential for improving the protective effect against dental caries, attention was focused upon the effect of fluoride on the developing tooth. Focus then moved towards studies of the role of fluoride in the complex interactions between the tooth tissues and their environment of plaque biofilm, saliva and pellicle. It is from these studies that many of the specific benefits of the role of fluoride in caries prevention have emerged.

While mechanisms behind fluoride-induced change to skeletal and dental tissues and the way fluoride behaves in protecting against dental caries are much clearer than they were 40 years ago, the area is still very complex and the plethora of literature is sometimes confusing.

This monograph has brought together current concepts relating to fluoride and its role in relation to the prevention of dental caries. Information from a large and complex field has been assembled in a clear sequence and presented in a very lucid fashion. Of particular note are the diagrams, which are very clear and a great help in presenting highly complex data in an easily understood context.

With this in mind, the text will be valuable for research workers or postgraduate students beginning a career in this or allied fields, and provide a clear up-to-date summary of current thinking in this area. Established researchers and teachers, whether in clinical or basic sciences, will also find the monograph a valuable addition to their libraries.

The value of this text stems from the contributions of distinguished researchers in this field. The editor, whose own laboratory has contributed substantially to this area in recent years, has brought together a number of internationally known authors with an impressive series of publications across the width of the fluoride research area.

With regard to the structure of the monograph, the first section deals with the availability of fluoride and how it is dealt with by the body from a physiological and metabolic standpoint. This forms the basis for and introduction to fluoride toxicity and the subject of fluorosis and the importance of monitoring intake. For the clinician, this highlights and clarifies the advantages of fluoride as well as possible hazards.

The second section focuses in more detail on modes of fluoride application and the way in which fluoride has been and is used to effect the dramatic reductions in dental caries with which it is associated. The complex mechanisms by which fluoride exerts its effects are described with clarity, and the entire text is accompanied by particularly useful illustrations.

Whether to those new to the field or to the established worker, this monograph will prove to be a most valuable resource to the field of fluoride research.

Colin Robinson, Leeds

Buzalaf MAR (ed): Fluoride and the Oral Environment.
Monogr Oral Sci. Basel, Karger, 2011, vol 22, pp 1–19

Fluoride Intake of Children: Considerations for Dental Caries and Dental Fluorosis

Marília Afonso Rabelo Buzalaf[a] · Steven Marc Levy[b]

[a]Department of Biological Sciences, Bauru Dental School, University of São Paulo, Bauru, Brazil;
[b]Departments of Preventive and Community Dentistry and Epidemiology, University of Iowa, Iowa City, Iowa, USA

Abstract

Caries incidence and prevalence have decreased significantly over the last few decades due to the widespread use of fluoride. However, an increase in the prevalence of dental fluorosis has been reported simultaneously in both fluoridated and non-fluoridated communities. Dental fluorosis occurs due to excessive fluoride intake during the critical period of tooth development. For the permanent maxillary central incisors, the window of maximum susceptibility to the occurrence of fluorosis is the first 3 years of life. Thus, during this time, a close monitoring of fluoride intake must be accomplished in order to avoid dental fluorosis. This review describes the main sources of fluoride intake that have been identified: fluoridated drinking water, fluoride toothpaste, dietary fluoride supplements and infant formulas. Recommendations on how to avoid excessive fluoride intake from these sources are also given.

Copyright © 2011 S. Karger AG, Basel

Fluorides play a key role in the prevention and control of dental caries. In the middle of the previous century, it was generally believed that fluoride had to be incorporated into dental enamel during development to exert its maximum protective effect. It was then considered unavoidable to have a certain prevalence and severity of fluorosis in a population to minimize the prevalence and severity of caries among children. In the 1980s, a paradigm shift regarding the cariostatic mechanisms of fluorides was proposed [1]. This considered that the predominant, if not entire, explanation of how fluorides control caries development is their topical effect on the de- and re-mineralization processes that occur at the interface between the tooth surface and the adjacent dental biofilm. This concept became widely accepted [2–6], and made it possible to obtain very substantial caries protection without significant ingestion of fluorides. With this in mind and being aware of the increase in the prevalence of dental fluorosis in both fluoridated and in non-fluoridated areas [7–9], researchers around the world turned their attention toward controlling the amount of fluoride intake.

The most important risk factor for fluorosis is the total amount of fluoride consumed from all sources during the critical period of tooth development. Thus, it is important not only to know the main sources of fluoride intake, but also the critical periods of formation in which the teeth are more susceptible to the effects of fluoride and the levels of fluoride intake above which dental fluorosis is expected to occur. The purpose of this review is to discuss the levels of fluoride intake that have been accepted as 'optimal' and the window of maximum susceptibility to the occurrence of dental fluorosis (focusing on the permanent maxillary central incisors), as well as to summarize the recent literature on risk factors for dental fluorosis, and describe the multiple sources of fluoride intake identified thus far and measures that should be adopted to reduce fluoride intake from these sources. All this information is of fundamental interest to clinicians who deal with children, in order that adequate counseling regarding fluoride intake can be provided to their parents.

'Optimal' Fluoride Intake

The widely accepted 'optimal' intake of fluoride (between 0.05 and 0.07 mg/kg) has been empirically established [10]. Its origin is attributed to McClure [11], who in the 1940s estimated that the 'average daily diet' contained 1.0–1.5 mg fluoride, which would provide about 0.05 mg/kg for children aged 1–12 years. Later on this information was interpreted as a recommendation when Farkas and Farkas [12] cited various sources that suggested 0.06 mg/kg fluoride was 'generally regarded as optimum'. In the 1980s, this range of estimates started being used as a recommendation for 'optimal' fluoride intake [13]. However, it is not clear if this level of intake is 'optimal' for caries prevention, for fluorosis prevention or a combination of both. It should also be noted that some authors regard 0.1 mg/kg per day to be the exposure level above which fluorosis occurs [14], although others have found dental fluorosis with a daily fluoride intake of less than 0.03 mg/kg per day [15]. It is worth mentioning that other factors may increase the susceptibility of individuals to dental fluorosis, including residence at high altitude [16–24], renal insufficiency [25–28], malnutrition [22, 29] and genetics [22, 30, 31]. Some of these factors can produce enamel changes that resemble dental fluorosis in the absence of significant exposure to fluoride (for details, see Buzalaf and Whitford, this vol., pp. 20–36).

Data from a recent cohort study (Iowa Fluoride Study) on longitudinal fluoride intake for children free of fluorosis in the early-erupting permanent dentition and free of dental caries in both the primary and early-erupting permanent teeth were compiled in an attempt to add scientific evidence to the 'optimal fluoride intake' [32]. The estimated mean daily fluoride intake for those children with no caries history and no fluorosis at age 9 years was at or below 0.05 mg/kg during different periods of the first 48 months of life, and this level declined thereafter. Children with caries generally had slightly lower intakes, whereas those with fluorosis had slightly higher intakes. Despite this being the only recent outcome-based assessment of 'optimal' fluoride intake, the overlap among caries/fluorosis groups in mean fluoride intake and the high variability in individual fluoride intakes for those with no fluoride or caries history discourage the strict recommendation of an 'optimal' fluoride intake. When it is necessary to employ parameters of 'optimal' fluoride intake, the range of 0.05–0.07 mg/kg should still be used.

Window of Maximum Susceptibility to the Development of Fluorosis

Considering that fluorotic changes in teeth cannot be reversed but may easily be prevented by controlling fluoride intake during the critical period of tooth formation, the identification of periods during which fluoride intake most strongly results in enamel fluorosis assumes great importance.

For the whole permanent dentition (excluding the third molars), the age for possible fluorosis development has been considered to be the first 6–8 years of life [33, 34]. However, most of the studies concerning the window of maximum susceptibility to dental fluorosis development have focused on the permanent maxillary central incisors, which are of the greatest cosmetic importance. While there is general consensus that the early maturation stage of enamel development is more critical

Table 1. Window of maximum susceptibility to the development of dental fluorosis in the permanent maxillary central incisors

Type of study	n	Window of maximum susceptibility	Fluoride source	References
1	86	6–23 months	toothpaste, supplements	Holm and Andersson [40], 1982
2	16	35–42 months	water	Ishii and Suckling [51], 1986
1	139	first 2 years	toothpaste	Osuji et al. [41], 1988
2	1,062	22–26 months	water	Evans and Stamm [38], 1991
2	1,085	15–24 months (males) 21–30 months (females)	water	Evans and Darvel [50], 1995
1	113	first 2 years	toothpaste	Lalumandier and Rozier [42], 1995
1	48	first year	water	Ismail and Messer [43], 1996
1	383	0–20 months	toothpaste, supplements	Wang et al. [44], 1997
1	66	first 2 years	water, toothpaste, supplements	Bårdsen and Bjorvatn [45], 1998
1 and 2[a]	n.a.	first 2 years (but duration of exposure more important)	variable	Bårdsen [52], 1999
2	1,896	first 3 years	water	Burt et al. [48], 2000 Burt et al. [49], 2003
1[b]	579	first 2 years	total intake	Hong et al. [46], 2006
1[b]	628	first 3 years	total intake	Hong et al. [47], 2006

Study type 1 = Individuals introduced to fluoride at different ages; study type 2 = populations exposed from birth that experienced an abrupt reduction in intake.
[a] Meta-analysis.
[b] Longitudinal design.

for fluorosis than the secretory stage [15, 35–39], the evidence is not completely conclusive regarding the age at which maxillary central incisors are most susceptible to dental fluorosis. The results of studies focused on this topic are summarized in table 1. They can be divided into two categories: studies involving subjects whose exposure to fluoride started at different ages during tooth formation [40–47] and those involving subjects that had

been exposed from birth and then had an abrupt reduction in daily fluoride intake [38, 48–51]. Most of these were cross-sectional, retrospective and focused on just one or two sources of fluoride intake. Only one more recent study used longitudinal data on individual fluoride intake [46, 47]. While one study reported that the first year of life was the most critical period for developing fluorosis in the permanent central maxillary incisors

[43], three studies found the first 3 years critical [47–49] and another recognized a later period (between 35 and 42 months) [51] – most of the studies agreed that the first 2 years of life are the most important [40–42, 44, 45] which was also the conclusion of a meta-analysis [52]. However, this meta-analysis acknowledged that the duration of fluoride exposure during amelogenesis, rather than specific risk periods, would seem to explain the development of dental fluorosis in the maxillary permanent central incisors, i.e. long periods of fluoride exposure (>2 out of the first 4 years) led to an odds ratio (OR) of 5.8 (95% CI 2.8–11.9) versus shorter periods of exposure (<2 out of the first 4 years of life). This is in line with data from a more recent longitudinal study which concluded that: (1) although the first 2 years of life were generally found to be more important compared with later years, fluoride intake during each individual year (until the fourth year of life) was associated with fluorosis; (2) subjects with higher levels of fluoride intake (estimated mean daily ingestion of 0.059 mg/kg) during the whole first 3 years of life had the highest risk of fluorosis [46]. Thus, the development of fluorosis appears to be related not only to the timing of fluoride intake relative to the periods of tooth formation, but also to the cumulative duration of fluoride exposure [46, 52].

From the available evidence, it seems rational to monitor fluoride intake of children in the first 3 years of life in order to minimize the risk of developing dental fluorosis of the permanent maxillary central incisors, which are the most relevant teeth from an aesthetic point of view [46, 47, 52].

Sources of Fluoride Intake

Concern with the increase in the prevalence of mostly mild but also some moderate-to-severe dental fluorosis and its potential impact on quality of life has led investigators all over the world to estimate the fluoride concentration of potential sources, as well as the fluoride intake from all sources – especially in children [53–82]. Case-control studies, cohort studies and randomized clinical trials whose results were compiled in systematic reviews with or without meta-analysis led to the identification of 4 major risk factors for dental fluorosis: fluoridated drinking water [83–85], fluoride supplements [86], fluoride toothpaste [87] and infant formulas [84]. Some manufactured infant foods and drinks may also be important contributors to the total daily fluoride intake [72–75, 78, 88]. These major sources of fluoride intake, as well as recommendations on how to reduce fluoride intake from them, will be discussed in detail below.

Fluoridated Drinking Water

Fluoridation of community drinking water is recognized among the top 10 greatest public health achievements in the world in the last century [89]. Although other fluoride-containing products are available, water fluoridation remains the most equitable and cost-effective method of delivering fluoride to all members of most communities, regardless of age, income level or educational attainment. Additionally, there is some evidence that water fluoridation may reduce the oral health gap between social classes [85]. The mean estimated costs for water fluoridation are only about USD 0.72/year per person in the USA [90].

Early in the 1940s, it was known that about 10% of children in areas naturally fluoridated at optimum levels (1.0 ppm) were affected by mild or very mild fluorosis of the permanent teeth, and this rate was less than 1% in low-fluoride areas [91]. These levels of prevalence were recorded when fluoridated drinking water was the only significant source of fluoride intake, before the widespread distribution of packaged beverages or the availability of fluoridated dental products. Studies conducted in the 1980s and 1990s reported that the prevalence of dental fluorosis in areas where the water fluoride concentration was 0.8 ppm was 4 times as high as that found in non-fluoridated communities [92–94]. In a systematic review of 214 studies on water fluoridation published in 2000, 88 studies on dental fluorosis were included [83]. The authors found a significant dose-response association between the fluoride concentration in the drinking water and the prevalence of dental fluorosis. It was estimated that, at a fluoride level of 1 ppm in the drinking water, the prevalence of any dental fluorosis was 48%, and 12.5% of exposed people had dental fluorosis that they would find of aesthetic concern (moderate to severe). This is much higher than that reported by Dean et al. [91] in 1942, who found virtually no cases of moderate or severe fluorosis, although the results are not directly comparable since different case definitions were used.

The studies that took advantage of breaks in water fluoridation to assess dental fluorosis of different birth cohorts are of special interest when analyzing the impact of fluoridated water on the prevalence of dental fluorosis in a community. In this way, Burt et al. [48] evaluated the impact of an unplanned break of 11 months in water fluoridation, and concluded that the prevalence of dental fluorosis is affected by changes in fluoride exposure from drinking water. However, in a subsequent study [49], the prevalence of dental fluorosis remained stable, in spite of an expected increase in the next cohort due to the re-establishment of water fluoridation. Buzalaf et al. [95] analyzed the effect of a 7-year interruption in water fluoridation on the prevalence of dental fluorosis in a Brazilian city. The authors found a lower prevalence of dental fluorosis in the permanent maxillary central incisors of children who were 36, 27 and 18 months old when water fluoridation ceased when compared with children who were born 18 months after fluoridation was interrupted. When analyzed together, the results of these studies conducted in the 2000s suggest that the relative importance of fluoridated water on the prevalence of dental fluorosis in current populations might not be as great as it was when

the only significant source of fluoride was the water supply.

The increased prevalence of dental fluorosis found more recently indicates that some young children are ingesting fluoride from sources other than drinking water. One study estimated that approximately 2% of US schoolchildren would experience perceived aesthetic problems which could be attributed to the currently recommended levels of fluoride in drinking water [96]. The US Department of Health and Human Services has recently proposed a new recommendation on water fluoride levels that is 0.7 ppm fluoride for the entire nation [97] and replaces the 1962 US Public Health Service Drinking Water Standards which were based on ambient air temperature of geographic areas and ranged from 0.7 to 1.2 ppm fluoride. This guidance is based on several considerations that include: (1) scientific evidence related to effectiveness of water fluoridation on caries prevention and control across all age groups; (2) fluoride in drinking water as one of several available fluoride sources; (3) trends in the prevalence and severity of dental fluorosis; (4) current evidence that fluid intake in children does not increase with increases in ambient air temperature due to augmented use of air conditioning and more sedentary lifestyles [98].

A recent study estimated the total daily fluoride intake from different constituents of the diet and from dentifrice by 1- to 3-year-old children living in an optimally fluoridated area. Standard fluoride concentration dentifrice alone was responsible for, on average, 81.5% of the daily fluoride intake, while among the constituents of the diet, water and reconstituted milk were the most important contributors and were responsible for about 60% of the total contribution of the diet [68]. For 4- to 6-year-old children living in the same community, however, the impact of fluoride ingested from dentifrice was less, and water alone provided a mean of 34% of the estimated daily fluoride from the diet, which corresponds to about 0.014 mg/kg [69, 70]. Thus, since fluoride present in water contributes only a small portion of intake from the dietary constituents, fluoridated water probably has its greatest impact on dental fluorosis prevalence indirectly, through being used in the reconstitution of infant formulas and in the processing of other children's foods and beverages [10]. Taking into account the low risks and great benefits of public water fluoridation, as well as the levels of prevalence and especially the severity of dental fluorosis found today, this measure must be maintained in the areas where it already exists and extended to the areas where it is feasible to implement water fluoridation.

In order to minimize the possible impact of water fluoridation on dental fluorosis, some measures should be taken. One of them is external monitoring of water fluoridation by an independent assessor. This measure has been shown to be successful in improving the consistency of fluoridation [99] and ideally should be implemented wherever there is adjusted fluoridation, but at least in the communities where fluctuations in water fluoride levels commonly occur [100].

It is also important to advise that, for infants and small children receiving large quantities of reconstituted infant formula, water containing <0.5 ppm fluoride should be used. A recent meta-analysis found that a 1.0-ppm increase in the fluoride level in the water supply is associated with a 67% increase in the OR for dental fluorosis associated with infant formula [84]. Thus, bottled water with relatively low fluoride content can be used instead of fluoridated water from the public supply [73, 77, 101]. Many brands of bottled water commercially available have low fluoride content and should be adequate for this purpose [53, 102–108]. However, one difficulty is that in many cases fluoride concentrations are not stated or are stated inaccurately on the labels, and unexpectedly high fluoride concentrations can be found [102, 106]. This reinforces the need for global labeling of fluoride levels in bottled water and rigorous surveillance by the competent public health authorities.

Fluoride Toothpaste

For several decades, fluoridated water was recognized as the major risk factor for dental fluorosis as a result of the classic studies by Dean et al. [91]. Observations that the prevalence of dental fluorosis had increased more in non-fluoridated than in fluoridated areas [109] resulted in efforts to better understand the relative impact of other potential sources of fluoride ingestion on the prevalence of dental fluorosis. Among them, fluoride toothpastes were identified as a potential risk factor for dental fluorosis, since an inverse relationship can be observed between the age of the child and the mean percentage ingestion of toothpaste [110]. A recent review compiled data for the estimated total fluoride intake of children living in different locations [111]. It was noted that toothpaste was usually the main contributor for young children. Thus, toothpaste is an important source of fluoride during the critical period of tooth development.

Table 2 summarizes the main findings of cross-sectional, case-control, cohort studies and randomized clinical trials conducted in different countries, both in fluoridated and non-fluoridated communities, which investigated the association between the use of fluoride toothpaste and the prevalence or severity of dental fluorosis. A positive association was found in most of these studies, mainly related to the early use of fluoride toothpaste (before age 24 months), regardless of the community fluoridation status. This issue was addressed in a recent systematic review and meta-analysis [87] compiling the results of 25 studies published between 1988 and 2006 that investigated the relationship between the use of fluoride toothpastes and dental fluorosis. Among these, two RCTs [112, 113], one cohort study [114], six case-control studies [36, 41, 115–118] and sixteen cross-sectional surveys [44, 92, 93, 119–131] were included. Among the 25 studies included, only one RCT was considered at low risk of bias [112]. The main findings of this systematic review with meta-analysis were: (1) a

Table 2. Studies assessing the association between the use of fluoride toothpaste and dental fluorosis

Study design	n	Age, years	Country	Fluoride water or salt	Other risk factors	Main outcome related to fluoride dentifrice	References
Case-control	633	8–10	Canada	yes	infant formula	OR = 11.0 (brushing before 25 months)	Osuji et al. [41], 1988
Cross-sectional	556	6–12	USA	varied	water, rinses	no association	Szpunar and Burt [92], 1988
Case-control	850	11–14	USA	no	supplements, family income, infant formula	OR = 2.9	Pendrys and Katz [36], 1989
Cross-sectional	350	7.5	Australia	varied	water, weaning age	OR = 2.6	Riordan [93], 1993
Case-control	401	12–16	USA	yes	supplements, infant formula	OR = 2.80 (frequent brushing)	Pendrys et al. [116], 1994
RCT	1,523	9–10	Norway	varied	supplements	TF lower for children using 550-ppm fluoride dentifrice	Holt et al. [113], 1994
Case-control	157	8–17	USA	varied	water	higher risk of fluorosis in children who used larger amounts of dentifrice	Skotowski et al. [118], 1995
Case-control	708	5–19	USA	varied	supplements	OR = 3.0 (age when started brushing)	Lalumandier and Rozier [42], 1995
Case-control	460	10–13	USA	no	supplements	OR = 2.5 (early dentifrice use)	Pendrys et al. [117], 1996
Cross-sectional	383	8	Norway	no	supplements	use of dentifrice before 14 months increased prevalence of fluorosis	Wang et al. [44], 1997
Cross-sectional	325	8–9	UK	yes	not evaluated	fluoride ingestion from dentifrice associated with fluorosis	Rock and Sabieha [128], 1997
Cross-sectional	197	1–7	USA	yes	supplement use from ages 0 to 3 years	no association	Morgan et al. [131], 1998
Cross-sectional	1,189	12	India	no	not evaluated	OR = 1.83 (use of fluoride dentifrice before age 6 years); beginning brushing before age 2 years increased severity of fluorosis	Mascarenhas and Burt [124], 1998
Case-control	233	10–14	USA	yes	supplements (OR = 6.0 and 10.8 for early- and later-forming enamel surfaces, respectively); powdered formula (OR = 10.7 for later-forming enamel surfaces)	OR = 6.4 and 8.4 for early-and later-forming enamel surfaces, respectively (early use of fluoride dentifrice)	Pendrys and Katz [115], 1998

Table 2. Continued

Study design	n	Age, years	Country	Fluoride water or salt	Other risk factors	Main outcome related to fluoride dentifrice	References
Cross-sectional	752	7–8	Canada	no	formula feeding, supplements	no association	Brothwell and Limeback [121], 1999
Cross-sectional	3,500	7–14	USA	varied	supplements, continuous exposure to fluoride water	brushing before age 2 years increased risk of fluorosis	Kumar and Swango [166], 1999
Cross-sectional	314	11–12	Brazil	no	no association	OR = 4.4 (brushing before age 3 years)	Pereira et al. [126], 2000
Cross-sectional	763	10–14	USA	varied	non-fluoridated area: supplements; fluoridated area: supplements and powdered infant formula	early toothbrushing behaviors regardless of water fluoride levels	Pendrys [145], 2000
Cross-sectional	867	8–9	UK	varied	water	use of adult dentifrice	Tabari et al. [130], 2000
Cross-sectional	582	10	Australia	varied	fluorosis prevalence declined after reduction in use of supplements	fluorosis prevalence declined after use of low-fluoride dentifrices increased	Riordan [127], 2002
Cross-sectional	8,277	not informed	Canada	varied	supplements; high parental educational level	beginning brushing between 1st and 2nd birthdays increased fluorosis (vs. between 2nd and 3rd birthdays)	Maupomé et al. [125], 2003
Cross-sectional	4,128	11	Belgium	no	supplements: ever vs. never (OR = 1.31), taken not in milk vs. in milk (OR = 1.69); water fluoride concentration	toothbrushing frequency: ≥2/day vs. <2/day (OR = 1.4)	Bottenberg et al. [120], 2004
RCT	703	8–9	UK	no	not evaluated	all subjects identified with TF = 3 were found in the 1,450-ppm fluoride dentifrice group	Tavener et al. [167], 2004
Cross-sectional	320	6–9	Mexico	yes	main source of fluoride exposure: professionally vs. self-applied (OR = 2.13)	effect of supplementary sources different between children brushing before 2 years (OR = 6.15) and after (OR = 2.14)	Beltran-Valladares et al. [119], 2005
Cross-sectional	548	7–9	Sweden	no	no association	no association	Conway et al. [122], 2005

Table 2. Continued

Study design	n	Age, years	Country	Fluoride water or salt	Other risk factors	Main outcome related to fluoride dentifrice	References
Cohort	343	7–11	USA	varied	ingestion from beverages, selected foods and fluoride supplements at ages 16 months, 36 months and AUC ages 16–36 months	significant association between fluorosis and toothpaste ingestion at age 24 months	Franzman et al. [114], 2006
RCT	1,268	8–10	UK	no	prevalence of fluorosis significantly higher in less-deprived districts	prevalence of TF ≥2 and ≥3 significantly higher for groups receiving 1,450-ppm fluoride dentifrice vs. 440 ppm	Tavener et al. [112], 2006
Cross-sectional	1,373	6–12	Mexico	yes	salt fluoridation	toothbrushing frequency associated with fluorosis (OR = 1.63)	Vallejos-Sánchez et al. [168], 2006
Cross-sectional	699	12	Ireland and Germany	yes	no association	no association	Sagheri et al. [129], 2007
Cross-sectional	677	9–13	Australia	varied	fluoridated water	use of standard-concentration fluoridated dentifrice; eating/licking toothpaste were risk factors for fluorosis	Do and Spencer [123], 2007
Case-control	2,106	13	Norway	no	supplements: regular use and mild-to-moderate fluorosis (OR = 6.5)	no children who had exclusively used only a pea-sized amount of toothpaste (1,000 ppm fluoride) had mild-to-moderate fluorosis	Pendrys et al. [169], 2010

TF index = Thylstrup and Fejerskov index.

significant reduction in the risk of dental fluorosis was found if toothbrushing with fluoride toothpaste did not start until the age of 12 months, but the evidence for starting toothbrushing with fluoride toothpaste before the age of 24 months was inconsistent (data from case-control and cross-sectional studies); (2) no significant association was found between frequency of toothbrushing or amount of toothpaste used and fluorosis (data from cross-sectional surveys); (3) using toothpaste with a higher concentration of fluoride increased the risk of dental fluorosis (data from two RCTs; evidence from cross-sectional studies was inconsistent). From the available evidence, the authors concluded that decisions involving the use of topical fluorides (including toothpastes) should balance their benefits in caries prevention and the risk of causing dental fluorosis. They noted: 'if the risk of fluorosis is of concern, the fluoride level of toothpaste for young children is recommended to be lower than 1,000 ppm' [87]. Risk-benefit considerations are critical. A recent systematic review and meta-analysis of 83 independent trials concluded that only toothpastes containing ≥1,000 ppm fluoride have been proven to be beneficial for preventing caries in children and adolescents [132]. However, for the deciduous dentition (age related with the development of dental fluorosis), uncertainty regarding the effectiveness of low-fluoride toothpastes for preventing caries was reported due to the lack of trials [132]. An alternative to improve the anti-caries effectiveness of low-fluoride dentifrices might be pH reduction, since lowering the pH enhances the tendency for calcium fluoride formation on enamel [133]. A recent RCT evaluated the caries increment in high caries risk 4-year-old children living in a fluoridated area with the use of a low-fluoride (550 ppm) acidic (pH 4.5) liquid dentifrice. It was observed that the caries progression rate was similar to that found with the use of a conventional 1,100-ppm fluoride toothpaste [134]. Also, the long-term use of this acidic dentifrice was shown to result in lower fingernail fluoride concentrations of the children using this product compared to the control toothpaste [135], which supports lower fluoride intake. The tested formulation could be an alternative to standard fluoride concentration toothpaste in order to avoid dental fluorosis in young children, but additional clinical trials are necessary to provide unequivocal evidence on this matter. Also further work should be done in attempt to enhance the anti-caries efficacy of low-fluoride toothpastes in order to further maximize benefits and minimize risk of accidental ingestion.

In conclusion, based on the available evidence regarding the risks of caries and dental fluorosis, it seems reasonable to recommend low-fluoride (500 ppm) toothpastes for young children who are at risk of developing dental fluorosis in the permanent maxillary central incisors (<3 years of age) but have low caries risk, especially if they live in a fluoridated area. In all other cases, toothpastes containing at least 1,000 ppm fluoride should be used. Although to date there is not unequivocal evidence supporting the association between the amount of toothpaste used and dental fluorosis [87], it seems rational to recommend the use of a small amount of toothpaste by young children, which can be easily achieved using the 'transverse' [136] or 'drop' [137] techniques. It is equally important that young children brush under adult supervision and be instructed to expectorate the foam after toothbrushing as much as possible.

Dietary Fluoride Supplements

Table 2, which describes the studies that investigated the association between the use of fluoride toothpaste and the occurrence of dental fluorosis, also shows that the most cited risk factor for dental fluorosis besides fluoride toothpaste is fluoride supplements.

Dietary fluoride supplements were originally designed to help prevent dental caries in children living in fluoride-deficient areas. The recommended daily dose was based on the age of the child and fluoride concentration in the drinking water. In 1999, a systematic review of studies evaluating the association between the use of fluoride supplements by children living in non-fluoridated areas and dental fluorosis was carried out [138]. By conducting a Medline search between 1966 and 1997, the authors were able to perform a qualitative review of 10 cross-sectional/case-control studies [36, 42, 44, 94, 117, 139–143] and found a strong consistent association between the use of fluoride supplements and dental fluorosis. The meta-analysis of these studies estimated that the OR of dental fluorosis in children living in non-fluoridated areas who had regularly used supplements during the first 6 years of life when compared with non-users was about 2.5 [138]. Recently, the same group updated the former systematic review by including an additional 4 studies in the meta-analysis [115, 120, 144, 145]. This inclusion confirmed the positive association between the use of supplements and the occurrence of dental fluorosis. The OR for dental fluorosis increased by 84% for each year of fluoride supplement use between the ages of younger than 6 months and 7 years, but the first 3 years of life were considered more important [86]. It must be highlighted, however, that most cases of dental fluorosis attributable to the use of fluoride supplements are graded as mild, with little likelihood of causing social impact, despite there being relatively few studies on this latter issue [86, 146].

In the later systematic review, the authors also evaluated the effectiveness of fluoride supplements in preventing caries. They concluded that there is weak inconsistent evidence showing that fluoride supplements are effective at preventing caries in primary dentition. However, they are able to help prevent caries in the permanent teeth of school-aged children (>6 years) when used on a regular basis – primarily due to a topical effect [86].

From the available evidence regarding the associations of supplements with dental caries and dental fluorosis, it is clear that consideration of the risk-benefit ratio is necessary when prescribing supplements, as discussed earlier for fluoride toothpastes. There is consensus that fluoride supplements should not be prescribed in optimally fluoridated areas, for infants less than 6 months of age, nor for children who are at low risk of developing dental caries. Different policies, however, have been adopted by distinct countries and dental associations regarding the recommendations for the appropriate use of supplements to prevent caries. The conclusion of a workshop conducted in Australia in 2006 on the use of fluorides for caries prevention was that 'fluoride supplements in the form of drops or tablets to be chewed and/or swallowed should not be used' [147]. In the USA,

the American Dental Association recommends: (1) no supplements from birth to 6 months or for residents of areas containing more than 0.6 ppm fluoride in the drinking water; (2) 0.25 mg fluoride/day from 6 months to 3 years for children living in areas containing less than 0.3 ppm fluoride in the drinking water; (3) 0.50 and 0.25 mg/day for children aged 3–6 years, living in areas with less than 0.3 and 0.3–0.6 ppm fluoride in the drinking water, respectively, while double the dose is recommended from 6 to 16 years. It should be emphasized that the American Dental Association recommends dietary fluoride supplements should only be used for children who are at high risk of developing dental caries [148]. The Canadian Dental Association recommends supplements only for children who have high caries experience and whose total intake of fluoride is lower than 0.05–0.07 mg/kg [149]. This recommendation, however, is not practical because estimating the total fluoride intake from all sources is very difficult. A more practical view, recommended by a group of European experts in 1991, states that 'a dose of 0.5 mg/day fluoride should be prescribed for at-risk individuals from the age of 3 years' [150].

Considering the available evidence indicating that fluoride supplements only help prevent caries when regularly used by children older than 6 years of age, and that their use before this age (but especially during the first 3 years) is associated with dental fluorosis [86], the view of the Europeans seems to be the most rational one. However, for remote/special populations not receiving other fluoride and caries prevention measures, fluoride supplementation may also be appropriate.

Infant Formulas

Despite breastfeeding being recommended in health campaigns worldwide, in many cases it is impractical. Additionally, as infants are weaned from breast milk, most of them receive the majority of their nutrition from infant formula, especially in the first 4–6 months of life before they start receiving solid foods. Commercially prepared infant formulas are available as powder and liquid concentrates that have to be diluted with water before use, or as ready-to-feed formulations. While human milk [151] and cow's milk [152] have low fluoride concentrations (typically <0.01 and <0.10 ppm, respectively), this is not true for infant formulas that can have a high intrinsic fluoride content due to manufacturing procedures or an increased fluoride content due to the use of fluoridated water for reconstitution of powders or liquid concentrates [77].

Fluoride concentrations in infant formulas show wide variations when assessed in different countries. For infant powdered formulas marketed in Brazil, fluoride concentrations ranged from 0.01 to 0.75 ppm when reconstituted with deionized water, from 0.91 to 1.65 ppm when reconstituted with fluoridated drinking water (containing 0.9 ppm), and from 0.02 to 1.37 ppm when reconstituted with different brands of bottled mineral water [101]. In Australia, fluoride concentrations ranging from 0.03 to 0.53 ppm for powdered formulas prepared with non-fluoridated water were reported [77]. In Thailand and Japan, values ranging from 0.14 to 0.64 and 0.37 to 1.00 ppm, respectively, were found [60]. In Malaysia, fluoride concentrations ranging from 0.10 to 0.16 ppm when prepared with deionized water, and from 0.35 to 0.40 ppm when prepared with water containing 0.38 ppm, were observed for infant formulas [153]. In the USA, fluoride concentrations of ready-to-feed, concentrated liquid and powdered formulas prepared with deionized water were reported to be around 0.15, 0.12–0.27 and 0.13 ppm, respectively [53, 154], and these fluoride levels would result in an intake well below the upper limit of 0.10 mg/day established by the Institute of Medicine (Washington, D.C., USA) under normal consumption of formulas [154]. However, most infants are likely to exceed the upper tolerable limit if they are exclusively fed powdered infant formula reconstituted with 0.7–

1.0 ppm fluoridated water. Thus, it has been suggested that the intake of fluoride by infants from formulas is influenced more by the water used to reconstitute the formula than by the formulas themselves [53, 73, 101, 154, 156].

Soy-based infant formulas have been reported to have somewhat higher fluoride concentrations than milk-based ones [77, 101, 153, 154, 156]. A study conducted with Malaysian soy-based formulas found values ranging from 0.24 to 0.44 ppm when prepared with deionized water and from 0.45 to 0.47 ppm when prepared with fluoridated water (0.38 ppm) [153]. It has been reported that substantial consumption of some fluoride-rich soy-based infant formulas, even when reconstituted with deionized water, would provide a fluoride intake above the upper tolerable limit for 1-month-old children [77, 101, 156].

Considering the fluoride concentrations present in infant formulas themselves, as well as the concentrations of fluoride present in the water used to reconstitute them, the above-mentioned studies have considered infant formula consumption a potential risk factor for dental fluorosis. A recent systematic review attempted to clarify the association between use of infant formula from birth to age 24 months and dental fluorosis [84]. The authors compiled the results of 19 studies including 17,429 subjects with ages ranging from 2 to 17 years. Among these studies, one was a prospective cohort study [157], five were retrospective cohort studies [14, 21, 93, 143, 158], six were case-control studies [36, 41, 115–117, 159], four were cross-sectional studies [121, 160–162] and three were historical-control studies [48, 123, 163]. The summary OR from 17 studies relating infant formula use to dental fluorosis in the permanent dentition was 1.8 (95% CI 1.4–2.3), but there was significant heterogeneity in the magnitude of the OR among the studies, indicating that the summary OR must be interpreted with caution. A meta-regression provided weak evidence that the fluoride in the infant formula resulted in an increased risk of developing dental fluorosis.

However, the dental fluorosis risk associated with the use of infant formula depended on the level of fluoride in the water supply. An increase in the dental fluorosis OR of 5% was seen as the fluoride level of the water supply increased by 0.1 ppm (OR 1.05, 95% CI 1.02–1.09), such that a 1.0-ppm increase in the fluoride concentration in the water supply is associated with a 67% increased OR for dental fluorosis associated with infant formula (OR 1.67, 95% CI 1.18–2.36). The authors were not able to determine, however, whether liquid or powder infant formulas with or without reconstitution affected fluorosis risk differently, since only a few studies provided this information [84]. A recent cohort study (Iowa Fluoride Study) reported that greater fluoride intakes from reconstituted infant formulas at ages 3–9 months increased risk of mild dental fluorosis of the permanent maxillary incisors [164].

In summary, considering that the increased risk of dental fluorosis posed by the use of infant formulas depends mainly on the fluoride level of water supply [84] and that the reconstitution of formulas with 0.7–1.0 ppm fluoride may provide infants with a daily fluoride intake above that likely to cause some degree of dental fluorosis [154], it seems reasonable to advise that, for those receiving large quantities of reconstituted infant formula, water containing <0.5 ppm fluoride should be used for reconstitution. Bottled water with low fluoride concentrations could be used for this purpose [77, 101, 102]. However, fluoride concentrations both in infant formula and bottled water must be correctly displayed on product labels. Periodical analyses of fluoride concentrations present in infant formula and bottled water by government or private laboratories could contribute to ensure that the fluoride levels are adequately displayed on the labels.

Manufactured Infant Foods and Beverages
During infancy, important sources of fluoride include selected commercially available foods and beverages. Many studies have shown that the fluoride concentrations of infant foods and beverages span a wide range and depend mainly on the fluoride concentration in the water used to manufacture them [165].

Beikost is a collective term for foods other than milk or formula fed to infants. A wide variation (between 0.01 and 8.30 ppm) has been reported regarding the fluoride content of beikost. Chicken-based products usually present the highest values due to the inclusion of bones in the manufacturing process. In some studies, fish-based products also have been reported to have high fluoride content. In general, the fluoride concentrations of most beikost is usually low [165]. However, some cereals commonly added to milk that are usually consumed by infants in Brazil have been shown to have high fluoride concentrations. This is the case for Mucilon and Neston (Nestlé), which have fluoride concentrations of 2.4 and 6.2 ppm, respectively [72, 74]. A relatively high fluoride concentration was also found in ready-to-drink chocolate milk (1.2 ppm, Toddynho, Quaker) [72, 74].

It must be highlighted that fluoride present in the manufactured foods and beverages comes as a 'contaminant'. The manufacturers are usually unaware of these fluoride concentrations, which can vary among the different batches of products. Thus, market basket studies are important for determining the fluoride concentrations in manufactured foods and beverages, since there are no laws that require this information to be stated on the products' labels. Health professionals must be updated with respect to the available information, in order to adequately advise the parents of children at the age of risk for dental fluorosis. The general recommendation is that children under the age of 7 years, but mainly in the first 3 years of life, avoid substantial consumption of products with high-fluoride content, since they can significantly contribute to the total daily fluoride intake and increase the risk of dental fluorosis, especially when associated with other fluoride sources.

Conclusion

Improved understanding of the mechanisms of action of fluoride for caries prevention and control have made it possible to obtain substantial benefit from the use of fluoride with a minimum risk of side effects, since it was established that fluoride does not need to be ingested to be beneficial. This observation, concomitant with the reported increase in the prevalence of dental fluorosis, turned the attention of researchers worldwide toward the necessity of controlling fluoride intake. The most important risk factor for dental fluorosis is the total amount of fluoride consumed from all sources during the critical period of tooth development (the first 3 years of life for the permanent maxillary central incisors). The main sources of fluoride intake which are recognized as potential risk factors for developing dental fluorosis are high fluoride water, optimally fluoridated water, fluoride toothpaste, dietary fluoride supplements and infant formulas, especially if these are reconstituted with fluoridated water. Maintaining appropriate levels of fluoride in the water from the public supply, avoiding the ingestion of substantial quantities of optimally fluoridated water with reconstituted infant formulas, placing a small amount of fluoride toothpaste onto the toothbrush and supervising toothbrushing of pre-school children, as well as not routinely prescribing fluoride supplements for children at low risk of developing caries, those living in fluoridated areas and those below age 3 years (regardless of the status of community fluoridation) are the main measures recommended to reduce the total fluoride intake of young children during the period of greatest risk for developing dental fluorosis.

References

1 Fejerskov O, Thylstrup A, Larsen MJ: Rational use of fluorides in caries prevention: a concept based on possible cariostatic mechanisms. Acta Odontol Scand 1981;39:241–249.

2 ten Cate JM, Duijsters PP: Influence of fluoride in solution on tooth demineralization. I. Chemical data. Caries Res 1983;17:193–199.

3 ten Cate JM, Featherstone JD: Mechanistic aspects of the interactions between fluoride and dental enamel. Crit Rev Oral Biol Med 1991;2:283–296.

4 Øgaard B, Rølla G, Ruben J, Dijkman T, Arends J: Microradiographic study of demineralization of shark enamel in a human caries model. Scand J Dent Res 1988;96:209–211.

5 Øgaard B, Rølla G, Dijkman T, Ruben J, Arends J: Effect of fluoride mouthrinsing on caries lesion development in shark enamel: an in situ caries model study. Scand J Dent Res 1991;99:372–377.

6 Featherstone JD: The science and practice of caries prevention. J Am Dent Assoc 2000;131:887–899.

7 Khan A, Moola MH, Cleaton-Jones P: Global trends in dental fluorosis from 1980 to 2000: a systematic review. SADJ 2005;60:418–421.

8 Whelton HP, Ketley CE, McSweeney F, O'Mullane DM: A review of fluorosis in the European Union: prevalence, risk factors and aesthetic issues. Community Dent Oral Epidemiol 2004;32(suppl 1):9–18.

9 Clark DC: Trends in prevalence of dental fluorosis in North America. Community Dent Oral Epidemiol 1994;22:148–152.

10 Burt BA: The changing patterns of systemic fluoride intake. J Dent Res 1992;71:1228–1237.

11 McClure FJ: Ingestion of fluoride and dental caries: quantitative relations based on food and water requirements for children 1 to 12 years old. Am J Dis Child 1943;66:8.

12 Farkas CS, Farkas EJ: Potential effect of food processing on the fluoride content of infant foods. Sci Total Environ 1974;2:399–405.

13 Ophaug RH, Singer L, Harland BF: Estimated fluoride intake of average two-year-old children in four dietary regions of the United States. J Dent Res 1980;59:777–781.

14 Forsman B: Early supply of fluoride and enamel fluorosis. Scand J Dent Res 1977;85:22–30.

15 Baelum V, Fejerskov O, Manji F, Larsen MJ: Daily dose of fluoride and dental fluorosis. Tandlaegebladet 1987;91:452–456.

16 Angmar-Månsson B, Whitford GM: Environmental and physiological factors affecting dental fluorosis. J Dent Res 1990;69:706–713, discussion 21.

17 Irigoyen ME, Molina N, Luengas I: Prevalence and severity of dental fluorosis in a Mexican community with above-optimal fluoride concentration in drinking water. Community Dent Oral Epidemiol 1995;23:243–245.

18 Mabelya L, König KG, van Palenstein Helderman WH: Dental fluorosis, altitude, and associated dietary factors (short communication). Caries Res 1992;26:65–67.

19 Mabelya L, van Palenstein Helderman WH, van't Hof MA, Konig KG: Dental fluorosis and the use of a high fluoride-containing trona tenderizer (magadi). Community Dent Oral Epidemiol 1997;25:170–176.

20 Manji F, Baelum V, Fejerskov O, Gemert W: Enamel changes in two low-fluoride areas of Kenya. Caries Res 1986;20:371–380.

21 Rwenyonyi C, Bjorvatn K, Birkeland J, Haugejorden O: Altitude as a risk indicator of dental fluorosis in children residing in areas with 0.5 and 2.5 mg fluoride per litre in drinking water. Caries Res 1999;33:267–274.

22 Yoder KM, Mabelya L, Robison VA, Dunipace AJ, Brizendine EJ, Stookey GK: Severe dental fluorosis in a Tanzanian population consuming water with negligible fluoride concentration. Community Dent Oral Epidemiol 1998;26:382–393.

23 Akosu TJ, Zoakah AI: Risk factors associated with dental fluorosis in Central Plateau State, Nigeria. Community Dent Oral Epidemiol 2008;36:144–148.

24 Pontigo-Loyola AP, Islas-Marquez A, Loyola-Rodriguez JP, Maupome G, Marquez-Corona ML, Medina-Solis CE: Dental fluorosis in 12- and 15-year-olds at high altitudes in above-optimal fluoridated communities in Mexico. J Public Health Dent 2008;68:163–166.

25 Juncos LI, Donadio JV Jr: Renal failure and fluorosis. JAMA 1972;222:783–785.

26 Porcar C, Bronsoms J, Lopez-Bonet E, Valles M: Fluorosis, osteomalacia and pseudohyperparathyroidism in a patient with renal failure. Nephron 1998;79:234–235.

27 Turner CH, Owan I, Brizendine EJ, Zhang W, Wilson ME, Dunipace AJ: High fluoride intakes cause osteomalacia and diminished bone strength in rats with renal deficiency. Bone 1996;19:595–601.

28 Ibarra-Santana C, Ruiz-Rodriguez Mdel S, Fonseca-Leal Mdel P, Gutierrez-Cantu FJ, Pozos-Guillen Ade J: Enamel hypoplasia in children with renal disease in a fluoridated area. J Clin Pediatr Dent 2007;31:274–278.

29 Rugg-Gunn AJ, al-Mohammadi SM, Butler TJ: Effects of fluoride level in drinking water, nutritional status, and socioeconomic status on the prevalence of developmental defects of dental enamel in permanent teeth in Saudi 14-year-old boys. Caries Res 1997;31:259–267.

30 Everett ET, McHenry MA, Reynolds N, Eggertsson H, Sullivan J, Kantmann C, Martinez-Mier EA, Warrick JM, Stookey GK: Dental fluorosis: variability among different inbred mouse strains. J Dent Res 2002;81:794–798.

31 Huang H, Ba Y, Cui L, et al.: COL1A2 gene polymorphisms (Pvu II and Rsa I), serum calciotropic hormone levels, and dental fluorosis. Community Dent Oral Epidemiol 2008;36:517–522.

32 Warren JJ, Levy SM, Broffitt B, Cavanaugh JE, Kanellis MJ, Weber-Gasparoni K: Considerations on optimal fluoride intake using dental fluorosis and dental caries outcomes – a longitudinal study. J Public Health Dent 2009;69:111–115.

33 Pendrys DG: The fluorosis risk index: a method for investigating risk factors. J Public Health Dent 1990;50:291–298.

34 Pendrys DG: Analytical studies of enamel fluorosis: methodological considerations. Epidemiol Rev 1999;21:233–246.

35 Larsen MJ, Richards A, Fejerskov O: Development of dental fluorosis according to age at start of fluoride administration. Caries Res 1985;19:519–527.

36 Pendrys DG, Katz RV: Risk of enamel fluorosis associated with fluoride supplementation, infant formula, and fluoride dentifrice use. Am J Epidemiol 1989;130:1199–1208.

37 Pendrys DG, Morse DE: Use of fluoride supplementation by children living in fluoridated communities. ASDC J Dent Child 1990;57:343–347.

38 Evans RW, Stamm JW: An epidemiologic estimate of the critical period during which human maxillary central incisors are most susceptible to fluorosis. J Public Health Dent 1991;51:251–259.

39 Bronckers AL, Lyaruu DM, DenBesten PK: The impact of fluoride on ameloblasts and the mechanisms of enamel fluorosis. J Dent Res 2009;88:877–893.

40 Holm AK, Andersson R: Enamel mineralization disturbances in 12-year-old children with known early exposure to fluorides. Community Dent Oral Epidemiol 1982;10:335–339.

41 Osuji OO, Leake JL, Chipman ML, Nikiforuk G, Locker D, Levine N: Risk factors for dental fluorosis in a fluoridated community. J Dent Res 1988;67:1488–1492.

42 Lalumandier JA, Rozier RG: The prevalence and risk factors of fluorosis among patients in a pediatric dental practice. Pediatr Dent 1995;17:19–25.

43 Ismail AI, Messer JG: The risk of fluorosis in students exposed to a higher than optimal concentration of fluoride in well water. J Public Health Dent 1996;56:22–27.

44 Wang NJ, Gropen AM, Øgaard B: Risk factors associated with fluorosis in a non-fluoridated population in Norway. Community Dent Oral Epidemiol 1997;25:396–401.

45 Bårdsen A, Bjorvatn K: Risk periods in the development of dental fluorosis. Clin Oral Investig 1998;2:155–160.

46 Hong L, Levy SM, Broffitt B, Warren JJ, Kanellis MJ, Wefel JS, Dawson D: Timing of fluoride intake in relation to development of fluorosis on maxillary central incisors. Community Dent Oral Epidemiol 2006;34:299–309.

47 Hong L, Levy SM, Warren JJ, Broffitt B, Cavanaugh J: Fluoride intake levels in relation to fluorosis development in permanent maxillary central incisors and first molars. Caries Res 2006;40:494–500.

48 Burt BA, Keels MA, Heller KE: The effects of a break in water fluoridation on the development of dental caries and fluorosis. J Dent Res 2000;79:761–769.

49 Burt BA, Keels MA, Heller KE: Fluorosis development in seven age cohorts after an 11-month break in water fluoridation. J Dent Res 2003;82:64–68.

50 Evans RW, Darvell BW: Refining the estimate of the critical period for susceptibility to enamel fluorosis in human maxillary central incisors. J Public Health Dent 1995;55:238–249.

51 Ishii T, Suckling G: The appearance of tooth enamel in children ingesting water with a high fluoride content for a limited period during early tooth development. J Dent Res 1986;65:974–977.

52 Bårdsen A: 'Risk periods' associated with the development of dental fluorosis in maxillary permanent central incisors: a meta-analysis. Acta Odontol Scand 1999;57:247–256.

53 Van Winkle S, Levy SM, Kiritsy MC, Heilman JR, Wefel JS, Marshall T: Water and formula fluoride concentrations: significance for infants fed formula. Pediatr Dent 1995;17:305–310.

54 Ophaug RH, Singer L, Harland BF: Dietary fluoride intake of 6-month and 2-year-old children in four dietary regions of the United States. Am J Clin Nutr 1985;42:701–707.

55 Levy SM, Maurice TJ, Jakobsen JR: Feeding patterns, water sources and fluoride exposures of infants and 1-year-olds. J Am Dent Assoc 1993;124:65–69.

56 Levy SM, Kohout FJ, Guha-Chowdhury N, Kiritsy MC, Heilman JR, Wefel JS: Infants' fluoride intake from drinking water alone, and from water added to formula, beverages, and food. J Dent Res 1995;74:1399–1407.

57 Levy SM, Kohout FJ, Kiritsy MC, Heilman JR, Wefel JS: Infants' fluoride ingestion from water, supplements and dentifrice. J Am Dent Assoc 1995;126:1625–1632.

58 Levy SM, Warren JJ, Davis CS, Kirchner HL, Kanellis MJ, Wefel JS: Patterns of fluoride intake from birth to 36 months. J Public Health Dent 2001;61:70–77.

59 Clovis J, Hargreaves JA: Fluoride intake from beverage consumption. Community Dent Oral Epidemiol 1988;16:11–15.

60 Chittaisong C, Koga H, Maki Y, Takaesu Y: Estimation of fluoride intake in relation to F, Ca, Mg and P contents in infant foods. Bull Tokyo Dent Coll 1995;36: 19–26.

61 Chowdhury NG, Brown RH, Shepherd MG: Fluoride intake of infants in New Zealand. J Dent Res 1990;69:1828–1833.

62 Guha-Chowdhury N, Drummond BK, Smillie AC: Total fluoride intake in children aged 3 to 4 years – a longitudinal study. J Dent Res 1996;75:1451–1457.

63 Kimura T, Morita M, Kinoshita T, Tsuneishi M, Akagi T, Yamashita F, Watanabe T: Fluoride intake from food and drink in Japanese children aged 1–6 years. Caries Res 2001;35: 47–49.

64 Lima YB, Cury JA: [Fluoride intake by children from water and dentifrice]. Rev Saude Publica 2001;35:576–581.

65 Paiva SM, Lima YB, Cury JA: Fluoride intake by Brazilian children from two communities with fluoridated water. Community Dent Oral Epidemiol 2003;31:184–191.

66 Phantumvanit P, Shinawatra V, Poshyachinda U, Phijaisanit P, Traisup C: Dietary fluoride intake of 4–6 month old infants in Bangkok. J Dent Assoc Thai 1987;37:226–231.

67 Pessan JP, Pin ML, Martinhon CC, de Silva SM, Granjeiro JM, Buzalaf MA: Analysis of fingernails and urine as biomarkers of fluoride exposure from dentifrice and varnish in 4- to 7-year-old children. Caries Res 2005;39:363–370.

68 de Almeida BS, da Silva Cardoso VE, Buzalaf MA: Fluoride ingestion from toothpaste and diet in 1- to 3-year-old Brazilian children. Community Dent Oral Epidemiol 2007;35:53–63.

69 Rodrigues MH, Leite AL, Arana A, Villena RS, Forte FD, Sampaio FC, Buzalaf MA: Dietary fluoride intake by children receiving different sources of systemic fluoride. J Dent Res 2009;88:142–145.

70 Buzalaf MAR, Rodrigues MHC, Pessan JP, Leite AL, Arana A, Villena RS, Forte FD, Sampaio FC: Biomarkers of fluoride in children exposed to different sources of systemic fluoride. J Dent Res 2011;90:215–219.

71 Miziara AP, Philippi ST, Levy FM, Buzalaf MA: Fluoride ingestion from food items and dentifrice in 2–6-year-old Brazilian children living in a fluoridated area using a semiquantitative food frequency questionnaire. Community Dent Oral Epidemiol 2009;37:305–315.

72 Buzalaf MA, Granjeiro JM, Duarte JL, Taga ML: Fluoride content of infant foods in Brazil and risk of dental fluorosis. ASDC J Dent Child 2002;69:196–200, 125–126.

73 Buzalaf MA, Damante CA, Trevizani LM, Granjeiro JM: Risk of fluorosis associated with infant formulas prepared with bottled water. J Dent Child (Chic) 2004;71:110–113.

74 Buzalaf MA, de Almeida BS, Cardoso VE, Olympio KP, Furlani Tde A: Total and acid-soluble fluoride content of infant cereals, beverages and biscuits from Brazil. Food Addit Contam 2004;21:210–215.

75 Buzalaf MA, Granjeiro JM, Cardoso VE, da Silva TL, Olympio KP: Fluorine content of several brands of chocolate bars and chocolate cookies found in Brazil. Pesqui Odontol Bras 2003;17:223–227.

76 Buzalaf MA, Pinto CS, Rodrigues MH, Levy FM, Borges AS, Furlani TA, da Silva Cardoso VE: Availability of fluoride from meals given to kindergarten children in Brazil. Community Dent Oral Epidemiol 2006;34:87–92.

77 Silva M, Reynolds EC: Fluoride content of infant formulae in Australia. Aust Dent J 1996;41:37–42.

78 Levy SM: An update on fluorides and fluorosis. J Can Dent Assoc 2003;69:286–291.

79 Levy SM, Kiritsy MC, Warren JJ: Sources of fluoride intake in children. J Public Health Dent 1995;55:39–52.

80 Heilman JR, Kiritsy MC, Levy SM, Wefel JS: Fluoride concentrations of infant foods. J Am Dent Assoc 1997;128:857–863.

81 Heilman JR, Kiritsy MC, Levy SM, Wefel JS: Assessing fluoride levels of carbonated soft drinks. J Am Dent Assoc 1999;130:1593–1599.

82 Kiritsy MC, Levy SM, Warren JJ, Guha-Chowdhury N, Heilman JR, Marshall T: Assessing fluoride concentrations of juices and juice-flavored drinks. J Am Dent Assoc 1996;127:895–902.

83 McDonagh MS, Whiting PF, Wilson PM, Sutton AJ, Chestnutt I, Cooper J, Misso K, Bradley M, Treasure E, Kleijnen J: Systematic review of water fluoridation. BMJ 2000;321:855–859.

84 Hujoel PP, Zina LG, Moimaz SA, Cunha-Cruz J: Infant formula and enamel fluorosis: a systematic review. J Am Dent Assoc 2009;140:841–854.

85 Parnell C, Whelton H, O'Mullane D: Water fluoridation. Eur Arch Paediatr Dent 2009;10:141–148.

86 Ismail AI, Hasson H: Fluoride supplements, dental caries and fluorosis: a systematic review. J Am Dent Assoc 2008;139:1457–1468.

87 Wong MC, Glenny AM, Tsang BW, Lo EC, Worthington HV, Marinho VC: Topical fluoride as a cause of dental fluorosis in children. Cochrane Database Syst Rev 2010:CD007693.

88 Cardoso VE, Olympio KP, Granjeiro JM, Buzalaf MA: Fluoride content of several breakfast cereals and snacks found in Brazil. J Appl Oral Sci 2003;11:306–310.

89 CDC: Achievements in public health. 1900–1999: fluoridation of drinking water to prevent dental caries. Morb Mort Wkly Rep 1999;48:933–940.

90 Jones S, Burt BA, Petersen PE, Lennon MA: The effective use of fluoride in public health. Bull WHO 2005;83:670–676.

91 Dean HT, Arnold FA, Elvolve E: Additional studies of the relation of fluoride domestic waters to dental caries experience in 4,425 white children aged 12–14 years in 13 cities in 4 states. Public Health Rep 1942;57:1155–1179.

92 Szpunar SM, Burt BA: Dental caries, fluorosis, and fluoride exposure in Michigan schoolchildren. J Dent Res 1988;67:802–806.

93 Riordan PJ: Dental fluorosis, dental caries and fluoride exposure among 7-year-olds. Caries Res 1993;27:71–77.

94 Riordan PJ, Banks JA: Dental fluorosis and fluoride exposure in Western Australia. J Dent Res 1991;70:1022–1028.

95 Buzalaf MA, de Almeida BS, Olympio KP, da SCVE, de CSPSH: Enamel fluorosis prevalence after a 7-year interruption in water fluoridation in Jaú, Sao Paulo, Brazil. J Public Health Dent 2004;64:205–208.

96 Griffin SO, Beltran ED, Lockwood SA, Barker LK: Esthetically objectionable fluorosis attributable to water fluoridation. Community Dent Oral Epidemiol 2002;30:199–209.

97 U.S. Department of Health and Human Services: Proposed HHS Recommendation for Fluoride Concentration in Drinking Water for Prevention of Dental Caries. U.S. Federal Register 2011;76:2383–2388.

98 Sohn W, Heller KE, Burt BA: Fluid consumption related to climate among children in the United States. J Public Health Dent 2001;61:99–106.

99 Ramires I, Maia LP, Rigolizzo Ddos S, Lauris JR, Buzalaf MA: [External control over the fluoridation of the public water supply in Bauru, SP, Brazil]. Rev Saude Publica 2006;40: 883–889.

100 Buzalaf MA, Granjeiro JM, Damante CA, Ornelas F: Fluctuations in public water fluoride level in Bauru, Brazil. J Public Health Dent 2002;62:173–176.

101 Buzalaf MA, Granjeiro JM, Damante CA, de Ornelas F: Fluoride content of infant formulas prepared with deionized, bottled mineral and fluoridated drinking water. ASDC J Dent Child 2001;68: 37–41, 10.

102 Ramires I, Grec RH, Cattan L, Moura PG, Lauris JR, Buzalaf MA: [Evaluation of the fluoride concentration and consumption of mineral water]. Rev Saude Publica 2004;38:459–465.

103 Grec RH, de Moura PG, Pessan JP, Ramires I, Costa B, Buzalaf MA: [Fluoride concentration in bottled water on the market in the municipality of Sao Paulo]. Rev Saude Publica 2008;42:154–157.

104 Quock RL, Chan JT: Fluoride content of bottled water and its implications for the general dentist. Gen Dent 2009;57: 29–33.

105 Cochrane NJ, Saranathan S, Morgan MV, Dashper SG: Fluoride content of still bottled water in Australia. Aust Dent J 2006;51:242–244.

106 Ahiropoulos V: Fluoride content of bottled waters available in Northern Greece. Int J Paediatr Dent 2006;16:111–116.

107 Martinez-Mier EA, Soto-Rojas AE, Buckley CM, Zero DT, Margineda J: Fluoride concentration of bottled water, tap water, and fluoridated salt from two communities in Mexico. Int Dent J 2005;55:93–99.

108 Zohouri FV, Maguire A, Moynihan PJ: Fluoride content of still bottled waters available in the North-East of England, UK. Br Dent J 2003;195:515–518, discussion 507.

109 Leverett D: Prevalence of dental fluorosis in fluoridated and nonfluoridated communities – a preliminary investigation. J Public Health Dent 1986;46:184–187.

110 Richards A, Banting DW: Fluoride toothpastes; in Fejerskov O, Ekstrand J, Burt BA, (eds): Fluoride in Dentistry, ed 2. Copenhagen, Munksgaard, 1996, pp 328–344.

111 Clarkson J, Watt RG, Rugg-Gunn AJ, Pitiphat W, Ettinger RL, Horowitz AM, Petersen PE, ten Cate JM, Vianna R, Ferrillo P, Gugushe TS, Siriphant P, Pine C, Buzalaf MA, Pessan JP, Levy S, Chankanka O, Maki Y, Postma TC, Villena RS, Wang WJ, MacEntee MI, Shinsho F, Cal E, Rudd RE, Schou L, Shin SC, Fox CH: Proceedings: 9th World Congress on Preventive Dentistry (WCPD): 'Community Participation and Global Alliances for Lifelong Oral Health for All,' Phuket, Thailand, September 7–10, 2009. Adv Dent Res 2010;22:2–30.

112 Tavener JA, Davies GM, Davies RM, Ellwood RP: The prevalence and severity of fluorosis in children who received toothpaste containing either 440 or 1,450 ppm F from the age of 12 months in deprived and less deprived communities. Caries Res 2006;40:66–72.

113 Holt RD, Morris CE, Winter GB, Downer MC: Enamel opacities and dental caries in children who used a low fluoride toothpaste between 2 and 5 years of age. Int Dent J 1994;44:331–341.

114 Franzman MR, Levy SM, Warren JJ, Broffitt B: Fluoride dentifrice ingestion and fluorosis of the permanent incisors. J Am Dent Assoc 2006;137:645–652.

115 Pendrys DG, Katz RV: Risk factors for enamel fluorosis in optimally fluoridated children born after the US manufacturers' decision to reduce the fluoride concentration of infant formula. Am J Epidemiol 1998;148:967–974.

116 Pendrys DG, Katz RV, Morse DE: Risk factors for enamel fluorosis in a fluoridated population. Am J Epidemiol 1994;140:461–471.

117 Pendrys DG, Katz RV, Morse DE: Risk factors for enamel fluorosis in a nonfluoridated population. Am J Epidemiol 1996;143:808–815.

118 Skotowski MC, Hunt RJ, Levy SM: Risk factors for dental fluorosis in pediatric dental patients. J Public Health Dent 1995;55:154–159.

119 Beltrán-Valladares PR, Cocom-Tun H, Casanova-Rosado JF, Vallejos-Sanchéz AA, Medina-Solis CE, Maupomé G: Prevalence of dental fluorosis and additional sources of exposure to fluoride as risk factors to dental fluorosis in schoolchildren of Campeche, Mexico (in Spanish). Rev Invest Clin 2005;57:532–539.

120 Bottenberg P, Declerck D, Ghidey W, Bogaerts K, Vanobbergen J, Martens L: Prevalence and determinants of enamel fluorosis in Flemish schoolchildren. Caries Res 2004;38:20–28.

121 Brothwell DJ, Limeback H: Fluorosis risk in grade 2 students residing in a rural area with widely varying natural fluoride. Community Dent Oral Epidemiol 1999;27:130–136.

122 Conway DI, MacPherson LM, Stephen KW, Gilmour WH, Petersson LG: Prevalence of dental fluorosis in children from non-water-fluoridated Halmstad, Sweden: fluoride toothpaste use in infancy. Acta Odontol Scand 2005;63:56–63.

123 Do LG, Spencer AJ: Decline in the prevalence of dental fluorosis among South Australian children. Community Dent Oral Epidemiol 2007;35:282–91.

124 Mascarenhas AK, Burt BA: Fluorosis risk from early exposure to fluoride toothpaste. Community Dent Oral Epidemiol 1998;26:241–248.

125 Maupomé G, Shulman JD, Clark DC, Levy SM: Socio-demographic features and fluoride technologies contributing to higher fluorosis scores in permanent teeth of Canadian children. Caries Res 2003;37:327–334.

126 Pereira AC, Da Cunha FL, Meneghim MC, Werner CW: Dental caries and fluorosis prevalence study in a nonfluoridated Brazilian community: trend analysis and toothpaste association. ASDC J Dent Child 2000;67:132–135, 83.

127 Riordan PJ: Dental fluorosis decline after changes to supplement and toothpaste regimens. Community Dent Oral Epidemiol 2002;30:233–240.

128 Rock WP, Sabieha AM: The relationship between reported toothpaste usage in infancy and fluorosis of permanent incisors. Br Dent J 1997;183:165–170.

129 Sagheri D, McLoughlin J, Clarkson JJ: The prevalence of dental fluorosis in relation to water or salt fluoridation and reported use of fluoride toothpaste in school-age children. Eur Arch Paediatr Dent 2007;8:62–68.

130 Tabari ED, Ellwood R, Rugg-Gunn AJ, Evans DJ, Davies RM: Dental fluorosis in permanent incisor teeth in relation to water fluoridation, social deprivation and toothpaste use in infancy. Br Dent J 2000;189:216–220.

131 Morgan L, Allred E, Tavares M, Bellinger D, Needleman H: Investigation of the possible associations between fluorosis, fluoride exposure, and childhood behavior problems. Pediatr Dent 1998;20: 244–252.

132 Walsh T, Worthington HV, Glenny AM, Appelbe P, Marinho VC, Shi X: Fluoride toothpastes of different concentrations for preventing dental caries in children and adolescents. Cochrane Database Syst Rev 2010:CD007868.

133 Petersson LG, Lodding A, Hakeberg M, Koch G: Fluorine profiles in human enamel after in vitro treatment with dentifrices of different compositions and acidities. Swed Dent J 1989;13:177–183.

134 Vilhena FV, Olympio KP, Lauris JR, Delbem AC, Buzalaf MA: Low-fluoride acidic dentifrice: a randomized clinical trial in a fluoridated area. Caries Res 2010;44:478–484.

135 Buzalaf MA, Vilhena FV, Iano FG, Grizzo L, Pessan JP, Sampaio FC, Oliveira RC: The effect of different fluoride concentrations and pH of dentifrices on plaque and nail fluoride levels in young children. Caries Res 2009;43:142–146.

136 Villena RS: An investigation of the transverse technique of dentifrice application to reduce the amount of fluoride dentifrice for young children. Pediatr Dent 2000;22:312–317.

137 Vilhena FV, Silva HM, Peres SH, Caldana Mde L, Buzalaf MA: The drop technique: a method to control the amount of fluoride dentifrice used by young children. Oral Health Prev Dent 2008;6:61–65.

138 Ismail AI, Bandekar RR: Fluoride supplements and fluorosis: a meta-analysis. Community Dent Oral Epidemiol 1999;27:48–56.

139 de Liefde B, Herbison GP: Prevalence of developmental defects of enamel and dental caries in New Zealand children receiving differing fluoride supplementation. Community Dent Oral Epidemiol 1985;13:164–167.

140 Granath L, Widenheim J, Birkhed D: Diagnosis of mild enamel fluorosis in permanent maxillary incisors using two scoring systems. Community Dent Oral Epidemiol 1985;13:273–276.

141 Bagramian RA, Narendran S, Ward M: Relationship of dental caries and fluorosis to fluoride supplement history in a non-fluoridated sample of schoolchildren. Adv Dent Res 1989;3:161–167.

142 Woolfolk MW, Faja BW, Bagramian RA: Relation of sources of systemic fluoride to prevalence of dental fluorosis. J Public Health Dent 1989;49:78–82.

143 Clark DC, Hann HJ, Williamson MF, Berkowitz J: Influence of exposure to various fluoride technologies on the prevalence of dental fluorosis. Community Dent Oral Epidemiol 1994;22:461–464.

144 Hiller KA, Wilfart G, Schmalz G: Developmental enamel defects in children with different fluoride supplementation – a follow-up study. Caries Res 1998;32:405–411.

145 Pendrys DG: Risk of enamel fluorosis in nonfluoridated and optimally fluoridated populations: considerations for the dental professional. J Am Dent Assoc 2000;131:746–755.

146 Chankanka O, Levy SM, Warren JJ, Chalmers JM: A literature review of aesthetic perceptions of dental fluorosis and relationships with psychosocial aspects/oral health-related quality of life. Community Dent Oral Epidemiol 2010;38:97–109.

147 Australian Research Centre for Population Oral Health: The use of fluorides in Australia: guidelines. Aust Dent J 2006;51:195–199.

148 Rozier RG, Adair S, Graham F, Iafolla T, Kingman A, Kohn W, Krol D, Levy S, Pollick H, Whitford G, Strock S, Frantsve-Hawley J, Aravamudhan K, Meyer DM: Evidence-based clinical recommendations on the prescription of dietary fluoride supplements for caries prevention. J Am Dent Assoc 2010;141:1480–1489

149 Swan E: Dietary fluoride supplement protocol for the new millennium. J Can Dent Assoc 2000;66:362, discussion 3.

150 Clarkson J: A European view of fluoride supplementation. Br Dent J 1992;172:357.

151 Koparal E, Ertugrul F, Oztekin K: Fluoride levels in breast milk and infant foods. J Clin Pediatr Dent 2000;24: 299–302.

152 Buzalaf MA, Pessan JP, Fukushima R, Dias A, Rosa HM: Fluoride content of UHT milks commercially available in Bauru, Brazil. J Appl Oral Sci 2006;14:38–42.

153 Latifah R, Razak IA: Fluoride levels in infant formulas. J Pedod 1989;13: 323–327.

154 Siew C, Strock S, Ristic H, Kang P, Chou HN, Chen JW, Frantsve-Hawley J, Meyer DM: Assessing a potential risk factor for enamel fluorosis: a preliminary evaluation of fluoride content in infant formulas. J Am Dent Assoc 2009;140:1228–1236.

155 Trautner K, Einwag J: Human plasma fluoride levels following intake of dentifrices containing aminefluoride or monofluorophosphate. Arch Oral Biol 1988;33:543–546.

156 McKnight-Hanes MC, Leverett DH, Adair SM, Shields CP: Fluoride content of infant formulas: soy-based formulas as a potential factor in dental fluorosis. Pediatr Dent 1988;10:189–194.

157 Hong L, Levy SM, Warren JJ, Dawson DV, Bergus GR, Wefel JS: Association of amoxicillin use during early childhood with developmental tooth enamel defects. Arch Pediatr Adolesc Med 2005;159:943–948.

158 Forsman B: Dental fluorosis and caries in high-fluoride districts in Sweden. Community Dent Oral Epidemiol 1974;2:132–148.

159 Villa AE, Guerrero S, Icaza G, Villalobos J, Anabalon M: Dental fluorosis in Chilean children: evaluation of risk factors. Community Dent Oral Epidemiol 1998;26:310–315.

160 Walton JL, Messer LB: Dental caries and fluorosis in breast-fed and bottle-fed children. Caries Res 1981;15:124–137.

161 Ericsson Y, Ribelius U: Increased fluoride ingestion by bottle-fed infants and its effect. Acta Paediatr Scand 1970;59:424–426.

162 van der Hoek W, Ekanayake L, Rajasooriyar L, Karunaratne R: Source of drinking water and other risk factors for dental fluorosis in Sri Lanka. Int J Environ Health Res 2003;13:285–293.

163 Larsen MJ, Senderovitz F, Kirkegaard E, Poulsen S, Fejerskov O: Dental fluorosis in the primary and the permanent dentition in fluoridated areas with consumption of either powdered milk or natural cow's milk. J Dent Res 1988;67:822–825.

164 Levy SM, Broffitt B, Marshall TA, Eichenberger-Gilmore JM, Warren JJ: Associations between fluorosis of permanent incisors and fluoride intake from infant formula, other dietary sources and dentifrice during early childhood. J Am Dent Assoc 2010;141:1190–1201.

165 Fomon SJ, Ekstrand J: Fluoride intake by infants. J Public Health Dent 1999;59:229–234.

166 Kumar JV, Swango PA: Fluoride exposure and dental fluorosis in Newburgh and Kingston, New York: policy implications. Community Dent Oral Epidemiol 1999;27:171–180.

167 Tavener JA, Davies GM, Davies RM, Ellwood RP: The prevalence and severity of fluorosis and other developmental defects of enamel in children who received free fluoride toothpaste containing either 440 or 1450 ppm F from the age of 12 months. Community Dent Health 2004;21:217–223.

168 Vallejos-Sanchéz AA, Medina-Solis CE, Casanova-Rosado JF, Maupomé G, Minaya-Sanchéz M, Perez-Olivares S: Dental fluorosis in cohorts born before, during, and after the national salt fluoridation program in a community in Mexico. Acta Odontol Scand 2006;64:209–213.

169 Pendrys DG, Haugejorden O, Bårdsen A, Wang NJ, Gustavsen F: The risk of enamel fluorosis and caries among Norwegian children: implications for Norway and the United States. J Am Dent Assoc 2010;141:401–414.

Marília Afonso Rabelo Buzalaf
Department of Biological Sciences
Bauru Dental School, University of São Paulo
Al. Octávio Pinheiro Brisolla, 9–75
Bauru-SP, 17012–901 (Brazil)
Tel. +55 14 3235 8346, E-Mail mbuzalaf@fob.usp.br

Buzalaf MAR (ed): Fluoride and the Oral Environment.
Monogr Oral Sci. Basel, Karger, 2011, vol 22, pp 20–36

Fluoride Metabolism

Marília Afonso Rabelo Buzalaf[a] · Gary Milton Whitford[b]

[a]Department of Biological Sciences, Bauru Dental School, University of São Paulo, Bauru, Brazil;
[b]Department of Oral Biology, Medical College of Georgia, Augusta, Ga., USA

Abstract

Knowledge of all aspects of fluoride metabolism is essential for comprehending the biological effects of this ion in humans as well as to drive the prevention (and treatment) of fluoride toxicity. Several aspects of fluoride metabolism – including gastric absorption, distribution and renal excretion – are pH-dependent because the coefficient of permeability of lipid bilayer membranes to hydrogen fluoride (HF) is 1 million times higher than that of F⁻. This means that fluoride readily crosses cell membranes as HF, in response to a pH gradient between adjacent body fluid compartments. After ingestion, plasma fluoride levels increase rapidly due to the rapid absorption from the stomach, an event that is pH-dependent and distinguishes fluoride from other halogens and most other substances. The majority of fluoride not absorbed from the stomach will be absorbed from the small intestine. In this case, absorption is not pH-dependent. Fluoride not absorbed will be excreted in feces. Peak plasma fluoride concentrations are reached within 20–60 min following ingestion. The levels start declining thereafter due to two main reasons: uptake in calcified tissues and excretion in urine. Plasma fluoride levels are not homeostatically regulated and vary according to the levels of intake, deposition in hard tissues and excretion of fluoride. Many factors can modify the metabolism and effects of fluoride in the organism, such as chronic and acute acid-base disturbances, hematocrit, altitude, physical activity, circadian rhythm and hormones, nutritional status, diet, and genetic predisposition. These will be discussed in detail in this review.

Copyright © 2011 S. Karger AG, Basel

Fluorine is a natural component of the biosphere. It is the thirteenth most abundant element in the earth's crust, constituting in the combined state around 0.065% by weight of the crust. Due to the small radius of the fluorine atom, its effective surface charge is the highest among all elements. As a consequence, fluorine is the most electronegative and reactive of all elements and hardly occurs in nature in its elemental form. Instead, it is found most frequently as inorganic fluoride that is widely distributed [1]. Besides its ubiquitous natural occurrence, widespread acceptance of the cariostatic properties of fluoride has led to its addition to systemic (such as water, salt, sugar, milk and supplements) and topical vehicles (such as toothpastes, gels, foams, mouth rinses and varnishes) which are widely employed for caries control [Buzalaf et al., this vol., pp. 97–114; Pessan et al., this vol., pp. 115–132; Sampaio and Levy, this vol., pp. 133–145]. It can be inferred therefore that the human organism is broadly exposed to fluoride. The main

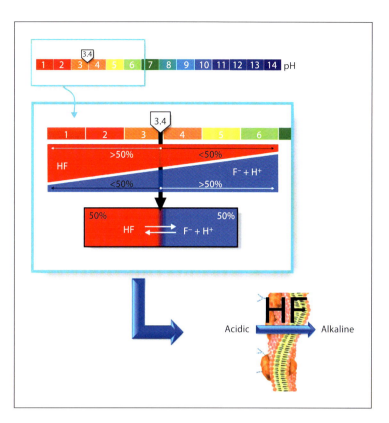

Fig. 1. pH-dependency of fluoride metabolism. HF is a weak acid with a pK_a of 3.4. Thus, at pH 3.4, 50% of fluoride is in the undissociated form (HF) while the remaining 50% is in the dissociated or ionic form (F^-). As pH decreases from 3.4, the concentration of HF increases, and as pH increases, the concentration of F^- increases. The coefficient of permeability of lipid bilayer membranes to HF is 1 million times higher than that of F^-. Therefore, fluoride crosses cell membranes as HF, in response to a pH gradient (goes from the more acidic compartment to the more alkaline compartment).

sources of fluoride intake were described in the chapter by Buzalaf and Levy [this vol., pp. 1–19]. Despite its proven benefits for caries control [2], there is a benefit/risk ratio that needs to be taken into account. The acute ingestion of a large dose can provoke gastric and kidney disturbances or even death in extreme cases [Whitford, this vol., pp. 66–80]. Lower levels of excessive intake on a chronic basis can affect the quality of the developing mineralized tissues, resulting in dental or skeletal fluorosis, depending on the amount and duration of intake [DenBesten and Li, this vol., pp. 81–96]. Thus, knowledge of all aspects of fluoride metabolism is essential for not only understanding the biological effects of this ion in humans, but also to optimize opportunities to prevent or treat cases of excess fluoride ingestion.

General Features of Fluoride Metabolism

Several aspects of fluoride metabolism – including gastric absorption, distribution and renal excretion – are pH-dependent. Hydrogen fluoride (HF) is a weak acid with a pK_a of 3.4. Thus, at pH 3.4, 50% of fluoride is in the undissociated form (HF) while the remaining 50% is in the dissociated or ionic form (F^-). As pH decreases from 3.4, the concentration of HF increases, and as pH increases, the concentration of F^- increases [3]. The coefficient of permeability of lipid bilayer membranes to HF is 1 million times higher than that of F^- [4]. This means that fluoride crosses cell membranes as HF, in response to a pH gradient between adjacent body fluid compartments, i.e. HF goes from the more acidic compartment to the more alkaline compartment (fig. 1).

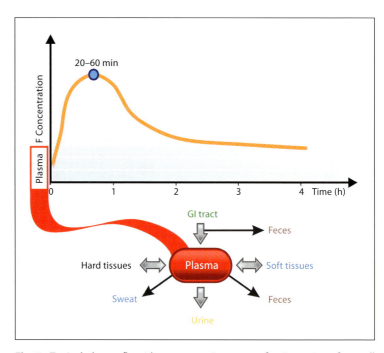

Fig. 2. Typical plasma fluoride concentration curve after ingestion of a small amount of fluoride and general features of fluoride metabolism. After ingestion, plasma fluoride levels increase rapidly, reaching a peak within 20–60 min due to absorption of fluoride in the GI tract and lung (to a lesser extent). Fluoride not absorbed will be excreted in feces. Plasma is the central compartment from which and into which fluoride must transit for its later distribution to hard and soft tissues and excretion. In adults, approximately 50% of an absorbed amount of fluoride will become associated with calcified tissues (mainly bone), where 99% of fluoride in the body is found. However, fluoride is not irreversibly bound to bone and can be released back into plasma when plasma fluoride levels fall (bidirectional arrows). A small amount of fluoride is found in soft tissues, where a steady-state distribution between extracellular and intracellular fluids is established. Most of the fluoride absorbed and not taken up by mineralized tissues is excreted in urine, while only a small amount of absorbed fluoride is excreted in sweat and feces.

General features of fluoride metabolism are described in figures 2 and 3. Figure 2 also illustrates a typical plasma fluoride concentration curve after ingestion of a small amount of fluoride. After ingestion, plasma fluoride levels increase rapidly (fig. 2) due to the ready absorption from the stomach, an event that is pH-dependent and distinguishes fluoride from other halogens and most other substances [3]. The majority of fluoride not absorbed from the stomach will be absorbed from the small intestine, but in this case absorption is not pH-dependent (fig. 3) [5, 6]. Fluoride not absorbed will be excreted in the feces [3].

Peak plasma fluoride concentrations are reached within 20–60 min following ingestion (fig. 2) and the levels start declining thereafter due to two main reasons: uptake in calcified

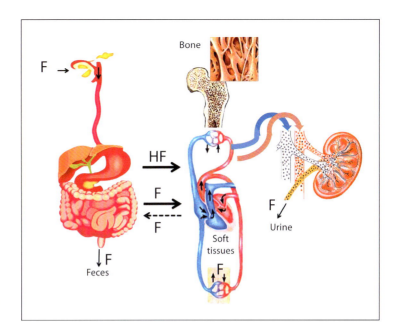

Fig. 3. General features of fluoride metabolism.

tissues and excretion in urine (fig. 2, 3). Plasma is the central compartment from which and into which fluoride must transit for its later distribution to hard and soft tissues and excretion. In adults, approximately 50% of an absorbed amount of fluoride will become associated with calcified tissues (mainly bone), where 99% of fluoride in the body is found [7]. However, fluoride is not irreversibly bound to bone and can be released back into plasma when plasma fluoride levels fall (bidirectional arrows in fig. 2, 3). A small amount of fluoride absorbed is found in soft tissues, where a steady-state distribution between extracellular and intracellular fluids is established. Most of the absorbed fluoride not taken up by mineralized tissues is excreted in urine while only a small amount of absorbed fluoride is excreted in sweat and feces. If the amount of fluoride ingested is small, the plasma fluoride concentrations return to baseline levels within 3–6 h (fig. 2) [3].

It is important to highlight that these general characteristics of fluoride metabolism are subject to variation due to dietary, environmental, genetic, physiological and pathological variables that will be discussed later in this chapter.

Fluoride Absorption

In the absence of high amounts of bi- and trivalent cations such as calcium, aluminum and magnesium that may complex fluoride and form insoluble compounds, approximately 80–90% of an amount of ingested fluoride is absorbed from the gastrointestinal tract [3]. Fluoride absorption occurs by passive diffusion (not against a concentration gradient), and is not affected by temperature changes or metabolic inhibitors. Fluoride absorption occurs rapidly, with a half time of approximately 30 min. Unlike most substances, roughly 20–25% of the total fluoride ingested is absorbed from the stomach, while the remainder is absorbed from the proximal small intestine [3, 6, 8, 9]. Although fluoride absorption from the stomach occurs rapidly, the rate of absorption is determined by gastric acidity [10, 11] and velocity of gastric emptying [6, 12]. Other factors that influence fluoride absorption are fluoride intake with other foods [13–15] and the specific salt of fluoride ingested [13, 15, 16].

Gastric fluoride absorption is inversely related to the pH of the stomach content because, in the stomach, fluoride is absorbed predominantly as HF [10]. When ionic fluoride enters the acidic gastric lumen environment, it is converted into HF which is an uncharged molecule that readily crosses cell membranes, including the gastric mucosa [4]. Thus, the higher the acidity of the gastric content, the faster the fluoride absorption from the stomach. As a consequence, peak plasma concentrations will be reached more quickly and sooner from an acidic environment than from a more neutral environment. The pH of the solution in which fluoride is administered, under conditions of normal gastric acid secretion, has little or no effect on fluoride absorption. However, animal studies have suggested that the pH of the solution exerts a profound short-term effect on fluoride absorption when drugs that inhibit gastric acid secretion are used. Solutions with lower pH would lead to a greater rate of fluoride absorption in the short term [11]. The extent of fluoride absorption from the stomach as a function of pH has important implications both for the treatment of acute fluoride toxicity [Whitford, this vol., pp. 66–80] and the therapeutic use of fluoride.

Another factor that interferes with gastric fluoride absorption is the rate of gastric emptying. Animal studies have shown that even at early time periods, while most of the fluoride dose still remained in the stomach, the majority of fluoride absorption occurred from the proximal small intestine. Thus, delayed gastric emptying might result in slower and smaller increases in plasma fluoride levels [6, 12].

Most of fluoride that is not absorbed from the stomach will be absorbed from the proximal small intestine (around 70–75% of absorbed fluoride) [5, 6]. The small intestine has a huge capacity for fluoride absorption and fluoride is rapidly absorbed following emptying from the stomach. Fluoride absorption from the small intestine, differently from what happens in the stomach, is unaffected by pH and occurs predominantly as the ionic fluoride (fig. 3) crosses the leaky epithelia through the tight junctions between the cells or paracellular channels [5]. The massive fluoride absorption from the small intestine compensates for the low gastric absorption at high pH, so that overall fluoride absorption is relatively unaffected by gastric acidity [11].

Fluoride absorption is affected by the composition of the diet and intake with foods. For a soluble fluoride compound, such as sodium fluoride (NaF) added to water, almost 100% of the fluoride is absorbed. If fluoride is ingested with milk (or baby formula) or with foods, especially those containing high amounts of divalent or trivalent cations that can complex fluoride and form insoluble compounds, the degree of absorption is reduced [13–15, 17]. This is the basis for using calcium-containing solutions to lavage the stomach in cases of acute fluoride toxicity [Whitford, this vol., pp. 66–80].

Regarding the type of fluoride ingested, most of the published studies are in agreement that the total amount of fluoride absorbed from disodium monofluorophosphate (SMFP) is similar to that absorbed from NaF [14, 16]. However, since absorption of fluoride from SMFP requires enzymatic hydrolysis of the moiety by phosphatases, fluoride absorption from SMFP occurs more slowly than from NaF. This leads to lower and delayed peak plasma fluoride levels compared to those seen after ingestion of NaF [13, 15, 16]. Similarly, the bioavailability of fluoride when ingested from naturally or artificially fluoridated water, which usually have different fluoride compounds, does not differ [18, 19].

Fluoride Distribution

After absorption, fluoride is rapidly distributed throughout the organism. Plasma fluoride levels start to increase within 10 min following fluoride intake and peak concentrations are reached within 20–60 min. Baseline plasma fluoride levels are usually reached within 3–11 h after ingestion, depending on the amount ingested [3].

From a pharmacokinetic point of view, plasma is regarded as the central compartment for fluoride distribution, since it is the fluid from which and into which fluoride must pass to be distributed to hard and soft tissues and excreted. A small part (<1%) of absorbed fluoride is found in soft tissues, where a steady-state distribution between extracellular and intracellular fluids is established [3]. This means that when there is an increase or decrease in plasma fluoride levels, a proportional change occurs in the fluoride concentrations of the extracellular and intracellular fluids. Most fluoride absorbed (around 35% for healthy adults) is taken up by calcified tissues where fluoride is reversibly bound and can be released back into plasma when plasma fluoride levels fall (fig. 2) [7].

The quantitative and qualitative aspects of fluoride distribution to each of these compartments will be detailed below.

Fluoride in Blood Plasma

There are two general forms of fluoride in human plasma. One fraction is ionic fluoride (also called inorganic or free fluoride) that can be detected by the fluoride ion-specific electrode. Ionic fluoride is not bound to other plasma constituents and is the form of significance in dentistry, medicine and public health. In the blood, ionic fluoride is not equally distributed between plasma and blood cells (its concentration in plasma is twice as high as that found in the cells). The other fraction is the non-ionic fluoride whose biological function has not been established yet, although its concentration is usually higher than that of ionic fluoride. This fraction: (1) seems to be composed chiefly of different types of lipid-like molecules that bind to plasma proteins; (2) can only be detected in plasma by the electrode after ashing; (3) is not expected to increase with increasing levels of chronic fluoride intake, suggesting little or no exchange between the two pools. Together, the non-ionic and ionic fractions constitute the so-called 'total' plasma fluoride [3, 20].

It is important to highlight that plasma ionic fluoride concentrations, unlike most other biologically relevant ions, are not homeostatically regulated. Instead they increase or decrease according to the amount of fluoride intake, deposition and removal in soft and hard tissues and urinary excretion [3]. As a consequence, plasma fluoride levels have been used as contemporary biomarkers of exposure to fluoride (indicate present exposure), although many physiological factors can influence plasma concentrations, regardless of fluoride intake [Rugg-Gunn et al., this vol., pp. 37–51].

Distribution to Soft Tissues

Fluoride in plasma is rapidly distributed to all tissues and organs. The velocity of distribution is determined by the rate of blood flow to the different tissues [20]. When considering fluoride distribution to soft tissues, it is useful to keep in mind that fluoride accumulates in the more alkaline compartment in response to a pH gradient (diffusion equilibrium of HF across cell membranes). In other words, fluoride goes from the more acidic to the more alkaline environment (fig. 1). Considering that the cytosol of mammalian cells is usually more acidic than extracellular fluid, intracellular fluoride levels are typically 10–50% lower than those found in plasma and extracellular fluid (fig. 4, 5), as shown by short-term experiments with radioactive fluoride in laboratory animals. However, intracellular fluoride concentrations change simultaneously and in proportion to changes in plasma fluoride levels [21]. Considering that the pH gradient across the membranes of most cells can be changed by

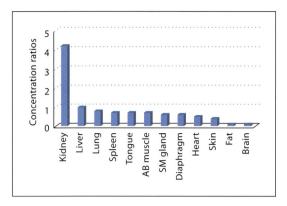

Fig. 4. Tissue/plasma fluoride (^{18}F) concentration ratios of soft tissues from the rat. AB = Abdominal; SM = submandibular [21].

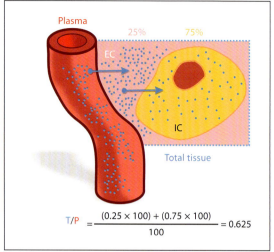

$$T/P = \frac{(0.25 \times 100) + (0.75 \times 100)}{100} = 0.625$$

Fig. 5. Distribution of fluoride in the water spaces of soft tissues. The concentrations of fluoride in plasma and interstitial fluid are assumed to be the same. The intracellular (IC) fluoride concentration is lower than that of the extracellular (EC) fluid. T/P = Tisue/plasma. Modified from Whitford [3].

altering extracellular pH, it is possible to promote the net flux of fluoride into or out of cells. For this reason, the recommended treatment in cases of acute and potentially toxic levels of fluoride ingestion includes alkalinization of the body fluids as a means to promote a net flux of fluoride out of cells, favoring fluoride elimination in the urine [22; Whitford, this vol., pp. 66–80].

Figure 4 shows tissue/plasma fluoride (^{18}F) concentration ratios of different soft tissues from published animal studies. The ratios are typically between 0.4 and 0.9 [21]. Exceptions are the brain (<0.1), because the blood-brain barrier is relatively impermeable to fluoride, and the kidney (>4.0), due to the high fluoride concentrations within the tubular and interstitial fluids.

Distribution to Specialized Body Fluids
Fluoride concentrations in some specialized body fluids are different from those found in plasma, but the concentrations change simultaneously and in proportion to those found in plasma. This is the case for cerebrospinal fluid and milk, which have fluoride concentrations 50% or less than that of plasma [3]. Gingival crevicular fluid fluoride levels are slightly higher than those in plasma, whereas the concentrations in parotid and submandibular ductal saliva are slightly lower. Ductal salivary-to-plasma fluoride concentration ratios have been reported to be around 0.9 and 0.8 for submandibular and parotid secretions, respectively [7]. Ductal saliva has been employed as a contemporary biomarker of fluoride exposure rather than plasma to estimate the bioavailability of fluoride from fluoridated products or fluoridated water [23–25]. Whole saliva usually has fluoride concentrations more variable and higher than those seen in ductal saliva due to exogenous contamination and is not recommended to estimate plasma fluoride levels [26]. For more details, see the chapter by Rugg-Gunn et al. [this vol., pp. 37–51].

Distribution to Mineralized Tissues
Fluoride is an avid mineralized tissue seeker. Approximately 99% of all fluoride retained in the human body is found in mineralized tissues, mainly in bone but also in enamel and dentin [7]. Fluoride

concentration in bone is not uniform. In long bone, the concentrations are higher in the periosteal and endosteal regions. Cancellous bone has higher fluoride concentrations than compact bone due to its greater surface area in contact with the surrounding extracellular fluid [27]. Bone fluoride concentrations tend to increase with age due to continuous fluoride uptake throughout life [27–29].

It is estimated that approximately 36% of the fluoride absorbed each day by healthy adults (18–75 years) becomes associated with the skeleton, while the remainder is excreted in urine. In children (<7 years), the degree of retention is much higher (around 55%) [30] due to the richer blood supply and larger surface area of bone crystallites, which are smaller, more loosely organized, and more numerous than those of mature bone [7].

Fluoride uptake by bone occurs in different stages [31]. The initial uptake occurs by iso- and heteroionic exchange on the hydration shells of bone crystallites. These ion-rich shells are continuous with the extracellular fluids. In fact, it is believed that a steady-state relationship exists between the fluoride concentrations in the extracellular fluids and the hydration shells of bone crystallites. According to this concept, there is a net transfer of fluoride from plasma to the hydration shells when the plasma concentration is rising and in the opposite direction when the plasma concentration is falling [7]. For this reason, bone surface has been suggested as a terminal biomarker of acute fluoride exposure [32–34; Rugg-Gunn et al., this vol., pp. 37–51]. Later stages involve fluoride association with or incorporation into precursors of hydroxyfluorapatite and finally into the apatitic lattice itself [31].

A physiologically based pharmacokinetic model considers that bone has two compartments: a small, flow-limited, rapidly exchangeable surface bone compartment and a bulk, virtually non-exchangeable, inner bone compartment. Fluoride associated with the inner bone compartment is not irreversibly bound. Over time, it may be mobilized through the continuous process of bone remodeling in the young, bone resorption and bone remodeling in the adult [35].

Dentin fluoride concentrations are similar to bone fluoride concentrations and both tend to increase with age, i.e. they are proportional to the long-term level of fluoride intake. Dentin fluoride levels are higher close to the pulp and reduce progressively towards the dentin-enamel junction [36]. Enamel fluoride concentrations are usually lower than the levels found in dentin; no correlation has been found between the fluoride concentrations in these two dental tissues [37, 38]. Enamel fluoride concentrations tend to decrease with age in areas subjected to tooth wear, but increase in areas that accumulate dental biofilm [39]. The fluoride concentrations of tooth enamel generally reflect the level of fluoride exposure during its formation [36]. However, a significant correlation between the severity of dental fluorosis and tooth fluoride concentrations has been found for dentin, but not for enamel [37, 38, 40].

Renal handling of Fluoride

Kidneys represent the major route of fluoride removal from the body. Under normal conditions, roughly 60% of fluoride absorbed each day by healthy adults (18–75 years) is excreted in urine. The corresponding percentage for children is 45% [30]. As a consequence, plasma and urinary excretion reflect a physiologic balance determined by previous fluoride intake, rate of fluoride uptake and removal from bone and the efficiency with which the kidneys excrete fluoride.

Since ionic fluoride is not bound to plasma proteins, its concentration in the glomerular filtrate is the same found in plasma. After entering the renal tubules, a variable amount of the ion is reabsorbed (from 10 to 90%) and returns to the systemic circulation, while the remainder in excreted in urine [20]. This process, together with glomerular filtration rate, is the main determinant of the amount of fluoride excreted in urine. The reduction in glomerular filtration rate that occurs in chronic renal dysfunction as well as in the last decades of life, when the number of functional nephrons is declining, will result in lower excretion and increased plasma fluoride levels [20, 41].

The renal clearance of fluoride (around 35 ml/ min in healthy adults) is unusually high when compared with the clearance of the other halogens (usually less than 1 or 2 ml/min). There is, however, a high variation among individuals [7] that is attributed to alterations in glomerular filtration rate [42], urinary pH [43–45] and flow rate [45, 46].

The mechanism of renal tubular reabsorption of fluoride, as happens for gastric absorption and transmembrane migration of fluoride, is also pH-dependent and occurs by diffusion of HF [44]. Thus, when the pH of the tubular fluid is relatively high, the proportion of fluoride as HF is lower while there is a higher proportion of fluoride as F^-. As a consequence, only a small amount of HF crosses the epithelium of the renal tubule to be reabsorbed and a high amount of fluoride is excreted in urine as F^-. On the other hand, when the pH of the tubular fluid is lower, high amounts of HF cross the tubular epithelium into the interstitial fluid where the pH is relatively high (around 7) which promotes the dissociation of HF. F^- is then released and diffuses into the peritubular capillaries returning to the systemic circulation. The renal clearance rate, in this case, is low (fig. 6). Thus, all conditions that alter urinary pH can affect the metabolic balance and tissue concentrations of fluoride. These include diet composition, certain drugs (such as ascorbic acid, ammonium chloride, chlorothiazide diuretics and methenamine mandelate), metabolic and respiratory disorders, and altitude of residence. These will be discussed later.

Fecal Fluoride

Most of the fluoride found in feces corresponds to the fraction that was not absorbed. Fecal fluoride usually accounts for less than 10% of the amount of ingested fluoride. Thus, more than 90% of ingested fluoride is usually absorbed [47, 48].

Fluoride present in feces, however, does not correspond solely to fluoride that was not absorbed. In two other situations, increased fecal fluoride excretion has been reported in rats: when plasma fluoride levels are high and when the diet contains high amounts of calcium (1% or higher). High plasma fluoride levels would cause net migration of fluoride from the systemic circulation into the intestinal tract. On the other hand, when diets containing high amounts of calcium are consumed, it is believed that unabsorbed calcium in the chyme binds fluoride entering the intestinal tract, thus reducing the concentration of diffusible fluoride and allowing the migration of more fluoride into the tract [49].

Factors That Modify the Metabolism or Effects of Fluoride

By analyzing the general features of fluoride metabolism, it becomes clear that any condition – systemic, metabolic or genetic – which interferes with the absorption or excretion of fluoride, will influence its fate in the body, and ultimately may alter the relationship between fluoride intake and the risk of dental or skeletal fluorosis. Variables that have been reported to modify the general features of fluoride metabolism in the organism include chronic and acute acid-base disturbances, hematocrit, high altitude, physical activity, circadian rhythm and hormones [3]. Other predisposing factors suggested are impaired kidney function, genetic predisposition and nutritional status. These will be discussed in more detail below.

Acid-Base Disturbances
Due to the effects of urinary pH on the efficiency of kidneys to remove fluoride from the body, chronic acid-base disturbances play an important role on the balance and tissue concentrations of fluoride. Factors that chronically alter the acid-

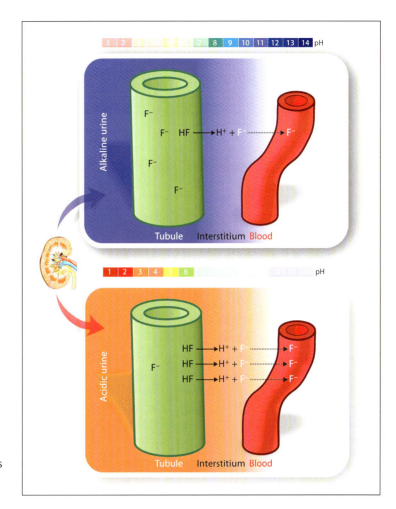

Fig. 6. Mechanism of fluoride reabsorption from the renal tubule. When urine is alkaline, there is a low concentration of HF and most of fluoride remains in the tubule to be excreted. When urine is acidic, there is a high concentration of HF that crosses the tubule membrane towards the interstitium where it dissociates originating F⁻ that diffuses into the peritubular capillaries and returns to the systemic circulation. Modified from Whitford [3].

base equilibrium include diet composition (vegetarian diet tends to increase urinary pH, while a diet with a high composition of meat tends to decrease urinary pH), certain drugs, a variety of metabolic and respiratory disorders, the level of physical activity and the altitude of residence [3]. Acute respiratory acid-base disorders affect renal excretion of fluoride in the same manner as the metabolic disorders [3].

Renal Impairment
Renal impairment in children has been associated with tooth defects that include enamel pitting and hypoplasia. The effects of uremia (increased concentrations of urea in blood) on tooth formation were evaluated in nephrectomized rats exposed to 0 or 50 ppm NaF in drinking water [50]. Intake of fluoride by nephrectomized rats increased plasma F levels twofold. It was also shown that uremia affected the formation of dentin and enamel, and was more extensive than the effect of fluoride alone, demonstrating that intake of fluoride by rats with reduced renal function impairs fluoride clearance from the plasma and aggravates the already negative effects of uremia on incisor tooth development. In humans, several studies have

shown a direct relationship between renal impairment and enamel defects, which include hypoplasia [51–53]. In a study comparing the frequency of dental fluorosis in children with renal disease and healthy children, although no significant difference was observed in the frequency of dental fluorosis between the 2 groups, patients with renal disease presented more severe dental fluorosis than children without renal disease [54].

Altitude of Residence
Researchers have noted that enamel disturbances are exacerbated in rats raised in hypobaric chambers which simulated high altitudes, regardless of the levels of ingested fluoride [3]. Alterations in acid-base balance, caused by hypobaric hypoxia during residence at high altitude, were cited as the cause of decreased urinary excretion of fluoride and therefore greater retention of fluoride [55]. In humans, a significantly higher prevalence of fluorosis has been observed in Tanzanian communities at a high altitude (1,463 m), in contrast with a low altitude area (100 m), but with similar food habits and low levels of fluoride in the drinking water [56]. The authors concluded that the severity of enamel disturbances at the high altitude area was not consistent with the low fluoride concentration in drinking water, suggesting that altitude, along with other factors, is a variable which may be contributing to the severity of dental enamel disturbances occurring in that area. Studies conducted in other countries confirmed this finding [57–60], suggesting that physiological changes associated with residence at high altitude are able to exacerbate the effects of fluoride in mineralized tissues. Such disturbances may be due to hypoxia in high altitude areas. This ultimately leads to a decrease in urinary pH, reducing fluoride renal excretion and, therefore, increasing fluoride concentrations in the body.

Physical Activity
In prolonged physical activity, there is a reduction in the pH gradient across cell membranes, especially skeletal muscle cells, which promotes the diffusion of fluoride (as HF) from the extracellular to the intracellular fluid. In addition, renal vasoconstriction can occur due to increased secretion of catecholamines and muscular blood flow during exercise. Depending on the balance of several factors, exercise could be associated with either decreased or increased circulating fluoride levels [3]. It must be considered, however, that although physical activity may alter the pattern of fluoride excretion, the impact of such findings on the development of dental fluorosis seem to be negligible, as prolonged physical activity in children at the age risk for fluorosis is uncommon.

Circadian Rhythm and Hormones
The possibility of existence of a biological rhythm in plasma fluoride levels was raised based on reports of circadian rhythms for calcium and phosphate [61, 62]. The daily variations of these ions are partially attributed to the balance between bone accretion and resorption, which are influenced by bone-active hormones. As the bulk of fluoride is contained in the skeleton, it was hypothesized that plasma fluoride levels would exhibit a circadian rhythm similar to, and in phase with, that of calcium and phosphate. Such rhythmicity was verified in dogs, with a mean peak fluoride concentration around 9 a.m., followed by a decrease around 9 p.m. [3].

The administration of parathormone or salmon thyrocalcitonin to humans demonstrated for the first time that alterations in hormone-mediated bone accretion and resorption are reflected in plasma and urinary fluoride levels. However, as reported in published animal studies [61, 62], the rhythmic pattern for calcium and phosphate occurred in the opposite way of that verified for fluoride, suggesting that a physiological system, other than bone, would be the responsible for the characteristics of the biological rhythmicity of fluoride in plasma. A recent study suggested that the renal system is involved with such rhythmicity in humans. Cardoso et al. [63] demonstrated a rhythmicity for fluoride concentrations in plasma, with mean peak (0.55 μmol/l) at 11 a.m. and the lowest concentrations (mean of 0.50 μmol/l) occurring between 5 and 8 p.m. Plasma fluoride concentrations were positively correlated with urinary fluoride excretion rates and with serum parathormone levels, suggesting that both the renal system and hormones might be involved in the rhythmicity for plasma fluoride concentrations in humans. It was also recently demonstrated that the diurnal average fractional urinary fluoride excretion is significantly lower than the average nocturnal one [64], which is in line with the findings of Cardoso et al. [63] and the suggested rhythmicity for plasma fluoride concentrations. The existence of this rhythmicity may alter the relationship between fluoride intake and the risk of dental or skeletal fluorosis.

Nutritional Status
Although an association between malnutrition and dental fluorosis prevalence and severity has been suggested for decades, the evidence for such a relationship is controversial and difficult to interpret. If a fasting child may absorb fluoride from water or other sources more quickly than a well-fed child (due to the inexistence of complexes of fluoride in an empty stomach), a malnourished child, on the other hand, may have low fluoride deposition over a long-term period of time (due to slower bone growth).

A statistically significant relationship between water fluoride concentration, socioeconomic status, nutritional status and the prevalence of diffuse enamel lesions (DDE index) in boys from Saudi Arabia has been demonstrated by Rugg-Gunn et al. [65]. Although the diffuse enamel defects of the DDE index are considered as an indicator of dental fluorosis, direct comparisons between the DDE index and specific dental fluorosis indices have been discouraged [66]. In a study with Tanzanian children, Yoder et al. [56] suggested a direct relationship between malnutrition and dental fluorosis. Such assumptions, however, must be considered with caution to avoid misinterpretation. The authors correlated their findings (high prevalence of dental fluorosis) with previous information on nutrition in 2 of the 3 areas evaluated, but no direct comparison between children with or without malnutrition regarding the prevalence of dental fluorosis was carried out.

Correia Sampaio et al. [67] demonstrated that dental fluorosis is independent of nutritional status. Nutritional status was assessed by the height-for-age (chronic malnutrition) and weight-for-age (general malnutrition) indexes, recommended by the WHO. A significant relationship between dental fluorosis and water fluoride concentration was found, but not with regard to nutritional status or sex. Dental fluorosis may be related to other factors, like infant dietary habits or increased consumption of fluoridated water. Future studies on this subject should consider a longitudinal study design where nutritional status, infant dietary habits and fluoride intake are assessed during the tooth formation period. This is particularly important for developing countries, where malnutrition and dental fluorosis are prevalent and fluoride-containing products are introduced in order to control dental caries.

Diet Composition
The acidification and subsequent alkalinization of urine by ingestion of NH_4Cl and $NaHCO_3$, respectively, led to significant differences in urinary fluoride clearance and plasma half-lives of 5 adult volunteers [68]. Similar findings were obtained by acidifying and alkalinizing urine by following a protein-rich (meat/dairy products) and a vegetarian diet, respectively. These results strongly suggested that long-term diet-induced changes in urinary pH could decrease (alkaline urine) or increase (acidic urine) the risk of dental fluorosis [55]. The prevalence and severity of dental fluorosis were compared among vegetarian and non-vegetarian children and adolescents living in an area with endemic dental fluorosis in India [69]. Vegetarianism was inversely associated with the

prevalence of dental fluorosis. The prevalence and severity (Thylstrup and Fejersko index ≥4) of dental fluorosis were 67 and 21%, respectively, in the vegetarian group, and 95 and 35%, respectively, in the nonvegetarian group (p < 0.05). In addition, multiple logistic regression analysis showed that the risk of developing dental fluorosis was 7 times higher among nonvegetarians than among vegetarians. Tamarind has also been shown to increase urinary fluoride excretion by increasing urinary pH in schoolchildren [70]. In a study conducted in a fluoride endemic area in South India, a significant decrease in urinary fluoride excretion was seen after volunteers were supplied with defluoridated water for 2 weeks. Then half of the subjects were supplemented with tamarind for 3 weeks, while the control group received defluoridated water for the same period. A significant increase in fluoride excretion and urinary pH was observed in the experimental group [71]. Tartaric acid is a major component of the tamarind paste (8.4–12.4%), which does not get metabolized and is excreted as such through the urine.

Other dietary constituents also seem to play an important role in the balance between fluoride and fluorosis. High dietary concentrations of certain cations, especially calcium, can reduce the extent of fluoride absorption [49]. In a study conducted in the province of Jiangxi, China, where the prevalence of dental fluorosis is reported to be above 50%, the incidence rates of dental fluorosis were found to differ markedly, depending on whether or not the children consumed milk. The rate of dental fluorosis of the milk-drinking group was 7.2%, whereas that of the non-milk-drinking group was 37.5% [72]. In India, where approximately 62 million people (including 6 million children) have dental fluorosis (mainly endemic), some studies have been conducted in order to identify components other than fluoride associated with an increased risk of dental fluorosis. Low calcium concentrations in the drinking water were demonstrated to be inversely related

to the prevalence of dental fluorosis [73, 74]. It was suggested that calcium supplementation should be implemented in areas with endemic fluorosis in order to minimize the effects of fluoride on mineralized tissues [74]. However, there is not enough evidence to support this, since none of the above-mentioned studies were able to determine the effect of calcium alone in communities with similar background exposure to fluoride from water.

The usual diet also appears to be important. Fluoride retention and resulting toxicity were found to be higher with sorghum (also called jowar) or sorghum-based diets than with rice- or wheat-based diets when the fluoride intakes were similar. Fluoride excretion in urine was significantly high on rice-based diets as compared with the sorghum-based diet [75].

Genetic Factors
Epidemiological observations of marked variation in dental fluorosis prevalence in subjects from areas with comparable levels of fluoride intake [56], or even in studies showing different degrees of susceptibility to fluorosis between certain ethnic groups [76–78] have led to the assumption that the predisposition to dental fluorosis is genetically determined [79, 80].

In a study conducted with Tanzanian children from three distinct areas, which differed regarding water fluoride concentrations and altitude, it was observed that even in the two sites with more severe fluorosis, several children had very little evidence of enamel disturbances [56]. These 'resistant' children were lifelong residents of the same area of the 'susceptible' children. Urinary fluoride values and meal fluoride values from children were also similar between the two groups. The question of possible genetic influence became more evident due to the tribal homogeneity in the area were fluorosis prevalence was unexpectedly high (no fluoridated drinking water).

The possibility of genetic predisposition to dental fluorosis was demonstrated using a mouse model system where genotype, age, gender, food, housing and drinking water fluoride levels were under control [81]. Examination of 12 inbred strains of mice showed differences in susceptibilities to dental fluorosis. The A/J mouse strain was highly susceptible, with a rapid onset and severe development of dental fluorosis compared to the other strains tested, whereas the 129P3/J mouse strain was less affected, with minimal dental fluorosis. It was latter demonstrated that these 2 strains also have different bone responses to fluoride exposure [82]. It was hypothesized that the different susceptibility to dental fluorosis between these two 2 strains was due to differences in fluoride metabolism, i.e. it was expected that the resistant strain would excrete more fluoride which in turn would lead to decreased susceptibility to dental fluorosis. Thus, a metabolic study was conducted to test this hypothesis. Surprisingly, the resistant strain (129P3/J) excreted a significantly lower amount of fluoride in urine than the susceptible strain (A/J) and, as a result, had significantly higher plasma and bone fluoride concentrations. Despite this, the amelogenesis in the 129P3/J strain was remarkably unaffected by fluoride [83]. Dental fluorosis-associated quantitative trait loci were detected on mouse chromosomes 2 and 11. Histological examination of maturing enamel showed that fluoride treatment resulted in accumulation of amelogenins in the maturing enamel of A/J mice, but not of 129P3/J mice [84]. The physiological, biochemical and/or molecular mechanisms underlying this resistance remain to be determined.

In humans, the possibility of gene-environment interaction was assessed by determining differential susceptibility to fluorosis at a given level of fluoride exposure based upon genetic background. A case-control study was conducted among children between 8 and 12 years of age with (n = 75) and without (n = 165) dental fluorosis in two counties in Henan Province, China. The PvuII and RsaI polymorphisms in the COL1A2

gene were genotyped. Calcitonin and osteocalcin levels in the serum were measured. Children carrying the homozygous genotype PP of COL1A2 PvuII had a significantly increased risk of dental fluorosis (OR = 4.85, 95% CI: 1.22–19.32) compared to children carrying the homozygous genotype pp in an endemic fluorosis village. However, the risk was not elevated when the control population was recruited from a non-endemic fluorosis village. Additionally, fluoride levels in urine and osteocalcin levels in serum were found to be significantly lower in controls from non-endemic villages compared to cases. However, the differences in fluoride and osteocalcin levels were not observed when cases were compared to a control population from endemic fluorosis villages. This study provided the first evidence of an association between polymorphisms in the COL1A2 gene with dental fluorosis in high-fluoride-exposed populations [85].

Conclusion

In view of the diverse effects that fluoride can produce in biological systems, it is not surprising that it has been the subject of thousands of scientific reports. It is clear that the beneficial as well as the adverse effects of fluoride can be attributed to the magnitude and duration of the concentration of the ion at specific tissue or cellular sites. In addition to the level of prior fluoride exposure, these concentrations are determined by the characteristics of the general metabolism of fluoride within the individual. As has been made clear in this chapter, these characteristics are not constant within or among individuals or populations. Instead they are subject to the effects of diverse environmental, biochemical, physiological and pathological factors. While much has been learned during the last few decades, much remains to be done – particularly in clearly defining the mechanisms underlying the metabolism and biological effects of fluoride. With the continuing development of advanced analytical, diagnostic, molecular and genetic techniques, we can expect our knowledge to grow and, with that growth, the beneficial effects of fluoride will be enhanced and the unwanted effects minimized.

Acknowledgments

The authors thank Prof. Heitor Marques Honório for designing figures 1, 2, 5 and 6 and Prof. Juliano Pelim Pessan for help with the section 'Factors that modify the metabolism or effects of fluoride'.

References

1 Smith FA, Ekstrand J: The occurrence and chemistry of fluoride; in Fejerskov O, Ekstrand J, Burt BA (eds): Fluoride in Dentistry, ed 2. Copenhagen, Munksgaard, 1996, pp 17–26.

2 Bratthall D, Hansel-Petersson G, Sundberg H: Reasons for the caries decline: what do the experts believe? Eur J Oral Sci 1996;104:416–422, discussion 423–425, 430–432.

3 Whitford GM: The metabolism and toxicity of fluoride. Monogr Oral Sci 1996;16:1–153.

4 Gutknecht J, Walter A: Hydrofluoric and nitric acid transport through lipid bilayer membranes. Biochim Biophys Acta 1981;644:153–156.

5 Nopakun J, Messer HH: Mechanism of fluoride absorption from the rat small intestine. Nutr Res 1990;10:771–9.

6 Nopakun J, Messer HH, Voller V: Fluoride absorption from the gastrointestinal tract of rats. J Nutr 1989;119:1411–1417.

7 Whitford GM: Intake and metabolism of fluoride. Adv Dent Res 1994;8:5–14.

8 Wagner MJ: Absorption of fluoride by the gastric mucosa in the rat. J Dent Res 1962;41:667–671.

9 Stookey GK, Dellinger EL, Muhler JC: In vitro studies concerning fluoride absorption. Proc Soc Exp Biol Med 1964;115:298–301.

10 Whitford GM, Pashley DH: Fluoride absorption: the influence of gastric acidity. Calcif Tissue Int 1984;36:302–307.

11 Messer HH, Ophaug RH: Influence of gastric acidity on fluoride absorption in rats. J Dent Res 1993;72:619–622.

12 Messer HH, Ophaug R: Effect of delayed gastric emptying on fluoride absorption in the rat. Biol Trace Elem Res 1991;31:305–315.

13 Ekstrand J, Ehrnebo M: Influence of milk products on fluoride bioavailability in man. Eur J Clin Pharmacol 1979;16:211–215.

14 Trautner K, Siebert G: An experimental study of bio-availability of fluoride from dietary sources in man. Arch Oral Biol 1986;31:223–228.

15 Trautner K, Einwag J: Influence of milk and food on fluoride bioavailability from NaF and Na_2FPO_3 in man. J Dent Res 1989;68:72–77.

16 Buzalaf MAR, Leite AL, Carvalho NTA, et al.: Bioavailability of fluoride administered as sodium fluoride or monofluorophosphate to humans. J Fluorine Chem 2008;129:691–694.

17 Spak CJ, Ekstrand J, Zylberstein D: Bioavailability of fluoride added by baby formula and milk. Caries Res 1982;16:249–256.

18 Maguire A, Zohouri FV, Mathers JC, Steen IN, Hindmarch PN, Moynihan PJ: Bioavailability of fluoride in drinking water: a human experimental study. J Dent Res 2005;84:989–993.

19 Whitford GM, Sampaio FC, Pinto CS, Maria AG, Cardoso VE, Buzalaf MA: Pharmacokinetics of ingested fluoride: lack of effect of chemical compound. Arch Oral Biol 2008;53:1037–1041.

20 Ekstrand J: Fluoride metabolism; in Fejerskov O, Ekstrand J, Burt BA (eds): Fluoride in Dentistry, ed 2. Copenhagen, Munksgaard, 1996, pp 55–68.

21 Whitford GM, Pashley DH, Reynolds KE: Fluoride tissue distribution: short-term kinetics. Am J Physiol 1979;236:F141–F148.

22 Whitford GM, Reynolds KE, Pashley DH: Acute fluoride toxicity: influence of metabolic alkalosis. Toxicol Appl Pharmacol 1979;50:31–39.

23 Zero DT, Raubertas RF, Fu J, Pedersen AM, Hayes AL, Featherstone JD: Fluoride concentrations in plaque, whole saliva, and ductal saliva after application of home-use topical fluorides [published erratum appears in J Dent Res 1993 Jan;72(1):87]. J Dent Res 1992;71:1768–1775.

24 Olympio KP, Bardal PA, Cardoso VE, Oliveira RC, Bastos JR, Buzalaf MA: Low-fluoride dentifrices with reduced pH: fluoride concentration in whole saliva and bioavailability. Caries Res 2007;41:365–370.

25 Wilson AC, Bawden JW: Salivary fluoride concentrations in children with various systemic fluoride exposures. Pediatr Dent 1991;13:103–105.

26 Whitford GM, Thomas JE, Adair SM: Fluoride in whole saliva, parotid ductal saliva and plasma in children. Arch Oral Biol 1999;44:785–788.

27 Weidmann SM, Weatherell JA: The uptake and distribution of fluorine in bones. J Pathol Bacteriol 1959;78:243–255.

28 Parkins FM, Tinanoff N, Moutinho M, Anstey MB, Waziri MH: Relationships of human plasma fluoride and bone fluoride to age. Calcif Tissue Res 1974;16:335–338.

29 Richards A, Mosekilde L, Sogaard CH: Normal age-related changes in fluoride content of vertebral trabecular bone – relation to bone quality. Bone 1994;15:21–26.

30 Villa A, Anabalon M, Zohouri V, Maguire A, Franco AM, Rugg-Gunn A: Relationships between fluoride intake, urinary fluoride excretion and fluoride retention in children and adults: an analysis of available data. Caries Res 2010;44:60–68.

31 Neuman WF, Neuman MW: The Chemical Dynamics of Bone Mineral, ed 1. Chicago, University of Chicago Press, 1958.

32 Bezerra de Menezes LM, Volpato MC, Rosalen PL, Cury JA: Bone as a biomarker of acute fluoride toxicity. Forensic Sci Int 2003;137:209–214.

33 Buzalaf MA, Caroselli EE, Cardoso de Oliveira R, Granjeiro JM, Whitford GM: Nail and bone surface as biomarkers for acute fluoride exposure in rats. J Anal Toxicol 2004;28:249–252.

34   Buzalaf MA, Caroselli EE, de Carvalho JG, de Oliveira RC, da Silva Cardoso VE, Whitford GM: Bone surface and whole bone as biomarkers for acute fluoride exposure. J Anal Toxicol 2005;29: 810–813.

35   Rao HV, Beliles RP, Whitford GM, Turner CH: A physiologically based pharmacokinetic model for fluoride uptake by bone. Regul Toxicol Pharmacol 1995;22:30–42.

36   Weatherell JA: Uptake and distribution of fluoride in bones and teeth and the development of fluorosis; in Barltrop W, Burland WL (eds): Mineral Metabolism in Paediatrics, ed 1. Oxford, Blackwell, 1969, pp 53–70.

37   Vieira AP, Hancock R, Dumitriu M, Limeback H, Grynpas MD: Fluoride's effect on human dentin ultrasound velocity (elastic modulus) and tubule size. Eur J Oral Sci 2006;114:83–88.

38   Vieira AP, Hancock R, Limeback H, Maia R, Grynpas MD: Is fluoride concentration in dentin and enamel a good indicator of dental fluorosis? J Dent Res 2004;83:76–80.

39   Weatherell JA, Robinson C, Hallsworth AS: Changes in the fluoride concentration of the labial enamel surface with age. Caries Res 1972;6:312–324.

40   Vieira A, Hancock R, Dumitriu M, Schwartz M, Limeback H, Grynpas M: How does fluoride affect dentin microhardness and mineralization? J Dent Res 2005;84:951–957.

41   Schiffl HH, Binswanger U: Human urinary fluoride excretion as influenced by renal functional impairment. Nephron 1980;26:69–72.

42   Spak CJ, Berg U, Ekstrand J: Renal clearance of fluoride in children and adolescents. Pediatrics 1985;75:575–579.

43   Jarnberg PO, Ekstrand J, Ehrnebo M: Renal excretion of fluoride during water diuresis and induced urinary pH-changes in man. Toxicol Lett 1983;18: 141–146.

44   Whitford GM, Pashley DH, Stringer GI: Fluoride renal clearance: a pH-dependent event. Am J Physiol 1976;230: 527–532.

45   Ekstrand J, Spak CJ, Ehrnebo M: Renal clearance of fluoride in a steady state condition in man: influence of urinary flow and pH changes by diet. Acta Pharmacol Toxicol (Copenh) 1982;50: 321–325.

46   Chen PS Jr, Gardner DE, Hodge HC, O'Brien JA, Smith FA: Renal clearance of fluoride. Proc Soc Exp Biol Med 1956;92:879–883.

47   Ekstrand J, Hardell LI, Spak CJ: Fluoride balance studies on infants in a 1-ppm-water-fluoride area. Caries Res 1984;18:87–92.

48   Ekstrand J, Ziegler EE, Nelson SE, Fomon SJ: Absorption and retention of dietary and supplemental fluoride by infants. Adv Dent Res 1994;8:175–180.

49   Whitford GM: Effects of plasma fluoride and dietary calcium concentrations on GI absorption and secretion of fluoride in the rat. Calcif Tissue Int 1994;54: 421–425.

50   Lyaruu DM, Bronckers AL, Santos F, Mathias R, DenBesten P: The effect of fluoride on enamel and dentin formation in the uremic rat incisor. Pediatr Nephrol 2008;23:1973–1979.

51   Koch MJ, Buhrer R, Pioch T, Scharer K: Enamel hypoplasia of primary teeth in chronic renal failure. Pediatr Nephrol 1999;13:68–72.

52   Al-Nowaiser A, Roberts GJ, Trompeter RS, Wilson M, Lucas VS: Oral health in children with chronic renal failure. Pediatr Nephrol 2003;18:39–45.

53   Farge P, Ranchin B, Cochat P: Four-year follow-up of oral health surveillance in renal transplant children. Pediatr Nephrol 2006;21:851–855.

54   Ibarra-Santana C, Ruiz-Rodriguez Mdel S, Fonseca-Leal Mdel P, Gutierrez-Cantu FJ, Pozos-Guillen Ade J: Enamel hypoplasia in children with renal disease in a fluoridated area. J Clin Pediatr Dent 2007;31:274–278.

55   Whitford GM: Determinants and mechanisms of enamel fluorosis. Ciba Found Symp 1997;205:226–241, discussion 241–245.

56   Yoder KM, Mabelya L, Robison VA, Dunipace AJ, Brizendine EJ, Stookey GK: Severe dental fluorosis in a Tanzanian population consuming water with negligible fluoride concentration. Community Dent Oral Epidemiol 1998;26: 382–393.

57   Rwenyonyi C, Bjorvatn K, Birkeland J, Haugejorden O: Altitude as a risk indicator of dental fluorosis in children residing in areas with 0.5 and 2.5 mg fluoride per litre in drinking water. Caries Res 1999;33:267–274.

58   Martinez-Mier EA, Soto-Rojas AE, Urena-Cirett JL, Katz BP, Stookey GK, Dunipace AJ: Dental fluorosis and altitude: a preliminary study. Oral Health Prev Dent 2004;2:39–48.

59   Pontigo-Loyola AP, Islas-Marquez A, Loyola-Rodriguez JP, Maupomé G, Marquez-Corona ML, Medina-Solis CE: Dental fluorosis in 12- and 15-year-olds at high altitudes in above-optimal fluoridated communities in Mexico. J Public Health Dent 2008;68:163–166.

60   Akosu TJ, Zoakah AI: Risk factors associated with dental fluorosis in Central Plateau State, Nigeria. Community Dent Oral Epidemiol 2008;36:144–148.

61   Talmage RV, Roycroft JH, Anderson JJ: Daily fluctuations in plasma calcium, phosphate, and their radionuclide concentrations in the rat. Calcif Tissue Res 1975;17:91–102.

62   Perault-Staub AM, Staub JF, Milhaud G: A new concept of plasma calcium homeostasis in the rat. Endocrinology 1974;95:480–484.

63   Cardoso VES, Whitford GM, Aoyama H, Buzalaf MAR: Daily variations in human plasma fluoride concentrations. J Fluorine Chem 2008;129:1193–1198.

64   Villa A, Anabalon M, Cabezas L, Rugg-Gunn A: Fractional urinary fluoride excretion of young female adults during the diurnal and nocturnal periods. Caries Res 2008;42:275–281.

65   Rugg-Gunn AJ, al-Mohammadi SM, Butler TJ: Effects of fluoride level in drinking water, nutritional status, and socioeconomic status on the prevalence of developmental defects of dental enamel in permanent teeth in Saudi 14-year-old boys. Caries Res 1997;31:259–267.

66   Clarkson J, O'Mullane D: A modified DDE Index for use in epidemiological studies of enamel defects. J Dent Res 1989;68:445–450.

67   Correia Sampaio F, Ramm von der Fehr F, Arneberg P, Petrucci Gigante D, Hatloy A: Dental fluorosis and nutritional status of 6- to 11-year-old children living in rural areas of Paraiba, Brazil. Caries Res 1999;33:66–73.

68   Ekstrand J, Ehrnebo M, Whitford GM, Jarnberg PO: Fluoride pharmacokinetics during acid-base balance changes in man. Eur J Clin Pharmacol 1980;18: 189–194.

69 Awadia AK, Haugejorden O, Bjorvatn K, Birkeland JM: Vegetarianism and dental fluorosis among children in a high fluoride area of northern Tanzania. Int J Paediatr Dent 1999;9:3–11.

70 Khandare AL, Rao GS, Lakshmaiah N: Effect of tamarind ingestion on fluoride excretion in humans. Eur J Clin Nutr 2002;56:82–85.

71 Khandare AL, Kumar PU, Shanker RG, Venkaiah K, Lakshmaiah N: Additional beneficial effect of tamarind ingestion over defluoridated water supply to adolescent boys in a fluorotic area. Nutrition 2004;20:433–436.

72 Chen YX, Lin MQ, Xiao YD, Gan WM, Min D, Chen C: Nutrition survey in dental fluorosis-afflicted areas. Fluoride 1997;30:77–80.

73 Bhargavi V, Khandare AL, Venkaiah K, Sarojini G: Mineral content of water and food in fluorotic villages and prevalence of dental fluorosis. Biol Trace Elem Res 2004;100:195–203.

74 Khandare AL, Harikumar R, Sivakumar B: Severe bone deformities in young children from vitamin D deficiency and fluorosis in Bihar-India. Calcif Tissue Int 2005;76:412–418.

75 Lakshmaiah N, Srikantia SG: Fluoride retention in humans on sorghum and rice based diets. Indian J Med Res 1977; 65:543–548.

76 Russell AL: Dental fluorosis in Grand Rapids during the seventeenth year of fluoridation. J Am Dent Assoc 1962; 65:608–612.

77 Butler WJ, Segreto V, Collins E: Prevalence of dental mottling in school-aged lifetime residents of 16 Texas communities. Am J Public Health 1985;75: 1408–1412.

78 Williams JE, Zwemer JD: Community water fluoride levels, preschool dietary patterns, and the occurrence of fluoride enamel opacities. J Public Health Dent 1990;50:276–281.

79 Anand JK, Roberts JT: Chronic fluorine poisoning in man: a review of literature in English (1946–1989) and indications for research. Biomed Pharmacother 1990;44:417–420.

80 Polzik EV, Zinger VE, Valova GA, Kazantsev VS, Yakusheva MY: A method for estimating individual predisposition to occupational fluorosis. Fluoride 1994; 27:194–200.

81 Everett ET, McHenry MA, Reynolds N, Eggertsson H, Sullivan J, Kantmann C, Martinez-Mier EA, Warrick JM, Stookey GK: Dental fluorosis: variability among different inbred mouse strains. J Dent Res 2002;81:794–798.

82 Mousny M, Banse X, Wise L, Everett ET, Hancock R, Vieth R, Devogelaer JP, Grynpas MD: The genetic influence on bone susceptibility to fluoride. Bone 2006;39:1283–1289.

83 Carvalho JG, Leite AL, Yan D, Everett ET, Whitford GM, Buzalaf MA: Influence of genetic background on fluoride metabolism in mice. J Dent Res 2009;88: 1054–1058.

84 Everett ET, Yan D, Weaver M, Liu L, Foroud T, Martinez-Mier EA: Detection of dental fluorosis-associated quantitative trait Loci on mouse chromosomes 2 and 11. Cells Tissues Organs 2009;189: 212–218.

85 Huang H, Ba Y, Cui L, et al.: COL1A2 gene polymorphisms (Pvu II and Rsa I), serum calciotropic hormone levels, and dental fluorosis. Community Dent Oral Epidemiol 2008;36:517–522.

Marília Afonso Rabelo Buzalaf
Department of Biological Sciences
Bauru Dental School, University of São Paulo
Al. Octávio Pinheiro Brisolla, 9–75
Bauru-SP, 17012–901 (Brazil)
Tel. +55 14 3235 8346, E-Mail mbuzalaf@fob.usp.br

Fluoride Intake, Metabolism and Toxicity

Buzalaf MAR (ed): Fluoride and the Oral Environment.
Monogr Oral Sci. Basel, Karger, 2011, vol 22, pp 37–51

Contemporary Biological Markers of Exposure to Fluoride

Andrew John Rugg-Gunn[a] · Alberto Enrique Villa[b] · Marília Afonso Rabelo Buzalaf[c]

[a]Newcastle University, Newcastle upon Tyne, UK; [b]Institute of Nutrition and Food Technology, University of Chile, Santiago, Chile; [c]Department of Biological Sciences, Bauru Dental School, University of São Paulo, Bauru, Brazil

Abstract

Contemporary biological markers assess present, or very recent, exposure to fluoride: fluoride concentrations in blood, bone surface, saliva, milk, sweat and urine have been considered. A number of studies relating fluoride concentration in plasma to fluoride dose have been published, but at present there are insufficient data on plasma fluoride concentrations across various age groups to determine the 'usual' concentrations. Although bone contains 99% of the body burden of fluoride, attention has focused on the bone surface as a potential marker of contemporary fluoride exposure. From rather limited data, the ratio surface-to-interior concentration of fluoride may be preferred to whole bone fluoride concentration. Fluoride concentrations in the parotid and submandibular/sublingual ductal saliva follow the plasma fluoride concentration, although at a lower concentration. At present, there are insufficient data to establish a normal range of fluoride concentrations in ductal saliva as a basis for recommending saliva as a marker of fluoride exposure. Sweat and human milk are unsuitable as markers of fluoride exposure. A proportion of ingested fluoride is excreted in urine. Plots of daily urinary fluoride excretion against total daily fluoride intake suggest that daily urinary fluoride excretion is suitable for predicting fluoride intake for groups of people, but not for individuals. While fluoride concentrations in plasma, saliva and urine have some ability to predict fluoride exposure, present data are insufficient to recommend utilizing fluoride concentrations in these body fluids as biomarkers of contemporary fluoride exposure for individuals. Daily fluoride excretion in urine can be considered a useful biomarker of contemporary fluoride exposure for groups of people, and normal values have been published.

Copyright © 2011 S. Karger AG, Basel

The concept of biological markers of fluoride exposure came to prominence in the 1994 Technical Report on Fluorides and Oral Health [1]. The WHO stated that 'a fluoride biomarker is of value primarily for identifying and monitoring deficient or excessive intakes of biologically available fluoride', and acknowledged a workshop on this topic, organized by the US National Institute of Dental Research, the year before [2]. In its report in 2002, the UK Medical Research Council [3] identified biomarkers of fluoride exposure as a research priority.

During the last two decades, there has been a rapid expansion in the availability and use of biomarkers in health care, such that they now occupy a central position in the armamentarium of the clinician for screening, diagnosis and management of disease [4]. For example, there is now extensive literature on biomarkers of bone formation [5], bone turnover [6] and bone resorption [7]. This expansion in bioinformatics has been due, in no small part, to rapid developments in chip technology in routine biological analysis [8]. However, people using biomarkers should be able to interpret results appropriately; for example, they should be able to answer the question 'Is it (the value obtained) normal?' The ideal method should be accurate, precise, sensitive and specific [9]. This chapter and the chapter by Pessan and Buzalaf [this vol., pp. 52–65] will review the literature on biological markers of fluoride exposure. As described below, this chapter will consider 'contemporary' biomarkers and the other will consider 'historical and recent' biomarkers of fluoride exposure.

Almost all fluoride entering the body is absorbed via the intestinal tract [Buzalaf and Whitford, this vol., pp. 20–36]. While most foods contain fluoride, fluoride is also ingested from fluoride-containing vehicles designed to control the development of dental caries. A proportion of fluoride entering the body is excreted in urine, and nearly all of the retained fluoride accumulates in calcified tissues, mainly bone. To maximize health benefits, there has to be a balance between too little fluoride (with increased risk of dental caries and its sequelae) and too much fluoride (with increased risk of fluorosis). The 'therapeutic ratio' is relatively low – the space between the two disbenefits (insufficiency and excess) is small. It is, therefore, very desirable to know what the body burden of fluoride is, in order to assess the risk/benefit ratio and maximize benefit and minimize disbenefit.

Buzalaf and Whitford [this vol., pp. 20–36] described how ingested fluoride was absorbed into the bloodstream, readily entering calcified tissue and excreted in urine. Thus, the body burden of fluoride might be estimated by examining fluoride concentrations in blood, bone, teeth and urine. In addition, fluoride concentrations in saliva, milk and sweat might reflect blood fluoride concentration, and fluoride concentration in nails (finger and toe) and hair might reflect past blood fluoride concentration and the body burden of fluoride. Fluoride toxicity can be 'acute' or 'chronic' – the former involving a very recent high or very high dose of fluoride, whilst the later might occur after modest yet excessive ingestion over a longer period of time. Biomarkers are needed for both situations – present exposure being assessed by 'contemporary' biomarkers, while more chronic fluoride exposure might be assessed by 'recent' or 'historic' biomarkers. Contemporary biomarkers might be fluoride concentrations in blood, bone surface, saliva, milk, sweat and urine, while historic biomarkers might include bone, teeth, nail and hair. This chapter will consider contemporary markers.

Early Investigations

The presence of fluoride in 'notable proportions' in the blood of humans and other animals was reported by Nickles in 1856 [10]. In 1888, Tammann recorded fluoride in the milk and blood of a cow and, in 1913, Gautier and Clausmann recorded the fluoride content of a large number of animal tissues including blood, milk and urine [10]. The first experimental study of fluoride deposition in soft tissues was reported in 1891 by Brandl and Tappeiner, who gave dogs daily food additives of 0.1–1.0 g sodium fluoride [10]. It is now known that the fluoride concentrations published were far too high due to inaccurate analytical methods [11], but these studies indicate that early investigators were aware of the possibility of fluoride toxicity. This lack of accuracy in fluoride analysis was particularly important for tissues such as blood and saliva where the fluoride concentration is low compared with mineralized tissues or urine [11]. Awareness grew during the 1930s and 1940s that water, along with sprayed pesticides and chemical contamination, were the principal potential sources of fluoride intake in man.

Controlled fluoride balance studies were conducted and, in a good summary, McClure et al. [12] stated that 'chronic cumulative toxic effects of fluorine may be predicted to a large extent by the quantity of fluorine which the body regularly retains'. They noted 'that fluorine concentration of spot urine specimens has a strikingly close correlation with the fluorine concentration in domestic water, through the range of 0.5 to 4.5 ppm fluorine in drinking water'. A crucial development occurred in the mid-1960s with the availability of the specific fluoride ion electrode which allowed accurate measurement of low concentrations of fluoride [11], although considerable care is needed for concentrations in the region of 0.01 mg/l (approximately 0.5 μmol/l) [13].

Blood

It is usual to report fluoride concentration in plasma, rather than in whole blood or serum. Blood cells contain about half the fluoride concentration recorded in plasma. Arguments for selecting plasma include: (1) plasma concentration establishes interstitial and intracellular fluoride concentrations in soft tissues, and (2) plasma is the fluid from which fluoride is filtered into the nephron [13]. While both ionic and non-ionic fluoride forms exist in plasma, the ionic form is of far greater significance [13], and it is detectable by the ion-specific electrode.

Plasma fluoride concentration returns to the resting value about 3–6 h after ingestion of a small fluoride dose; the half-life is about 30 min [13]. Fasting (baseline or resting) values are, therefore, usually determined after overnight fasting. There have been a number of reports of fasting plasma fluoride concentrations in humans [13–26], and these are listed in table 1, excluding determinations made before the availability of the ion-specific electrode. For less than optimal water fluoride concentrations, the resting fluoride concentrations ranged from 9.3 to 24.0 ng/ml (0.49–

1.26 μmol/l). Most of these studies reported the effect on plasma fluoride concentration of ingesting a dose of fluoride: the maximum plasma fluoride concentration was usually reported and these are also given in table 1. The effect of fluoride dose on plasma fluoride concentration is given in figure 1. The data points are taken from the mean values given in table 1: readers should be aware that this limits the interpretation of the plot, since the studies varied considerably in location, number and age of the subjects, and background fluoride exposure.

When interpreting plasma fluoride concentration information, it is important to be aware that several factors, independent of fluoride dose, influence the concentration value. These include: site of blood collection [13], age [13], acid-base balance [13], altitude [13], hematocrit [13] and genetic background [27]. The effects of circadian rhythm [13, 25] and hormones [13, 25] are rather discrete. Of these, site of blood collection, age and hematocrit have the most influence. It is advisable to collect venous or capillary blood [13]. Plasma fluoride concentration increases with age, and this is likely to be a reflection of increasing bone fluoride concentration with age [13]. Hematocrit values are likely to be lower in females than males [13].

Bone Surface

The human skeleton, which weighs approximately 3–4 kg, consists of 60% inorganic material and 30% organic matrix (95% type I collagen). The mineral phase consists chiefly of hydroxyapatite. Bone is metabolically active and about 5–10% of existing bone in adults is replaced through modeling each year. The skeleton consists of two types of bone: cortical bone (sometimes called compact bone) which comprises about 80% of the skeletal mass, and trabecular bone (also called spongy or cancellous bone) which forms the internal scaffolding of bone. Peak bone mass is achieved at

Table 1. Studies of plasma fluoride concentrations in human subjects

Study (first author)	Subjects			Resting		After fluoride dose (maximum)		
	n	age, years	water fluoride, ppm	ng/ml	µmol/l	fluoride dose, mg	ng/ml	µmol/l
Ekstrand [14]	1	27	0.25	10.3	0.54	10	300	15.8
Ekstrand [15]	5	24–28	0.25	10.0[1]	0.50			
	5	27–56	1.20	20.0[1]	1.00			
	5	10–38	9.60	35.0[1]	1.84			
Ekstrand [16]	5	27–36	n.a.	13.3	0.70[1]	1.5	75	4
Oliveby [17]	5	26–38	0.2	9.3	0.49[2]	1.0	51.1	2.69[2]
Oliveby [18]	5	26–38	0.2	9.5	0.50[3]	1.0	52.3	2.75[3]
Oliveby [19]	5	26–38	0.2	12.4	0.65[3]	1.0	52.6	2.77[3]
Whitford [13]	5	adults	n.a.	12.7	0.67	10.0	289	15.2
Whitford [20]	17	5–10	n.a.	16.9	0.89			
Levy [21]	15	2–6	0.6–0.8	19.0	1.00			
	15	2–6	0.1–0.2	24.0	1.26			
Maguire [22]	20	20–35	0.02	19.8	1.04	0.5	32.3[4]	1.7[4]
Cardoso [23]	5	25–35	0.03[5]	9.7	0.51			
	5	25–35	0.70	6.8	0.36[6]			
	5	25–35	0.30	10.5	0.55[7]			
Whitford [24]	5	24–32	0.85	17.3	0.91	0.33	28.3[8]	1.49[8]
	5	24–32	0.85	20.0	1.05	2.73	137.2[8]	7.22[8]
Cardoso [25]	5	27–33	low F[9]	10.0	0.53			
Buzalaf [26]	4	19–29	0.6–0.8[10]	21.0	1.11	2.00[11]	106.0	5.58
	4	19–29	0.6–0.8[10]	22.0	1.16	2.00[12]	100.0	5.26

[1] Estimated from graph.
[2] For study 1 and low flow rate.
[3] For unstimulated saliva.
[4] For subjects receiving naturally fluoridated soft water.
[5] Subjects drank bottled water.
[6] Low fluoride intake from dentifrice.
[7] High fluoride intake from dentifrice.
[8] For subjects receiving naturally fluoridated water.
[9] Subjects drank bottled water with low fluoride content.
[10] Personal communication from Buzalaf.
[11] Fluoride administered as sodium fluoride.
[12] Fluoride administered as disodium monofluorophosphate.

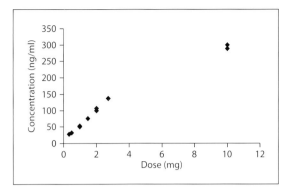

Fig. 1. Plot of the maximum concentration of fluoride in plasma after ingestion against the dose. Data are the mean values obtained from the 11 studies listed in table 1 (right-most columns). Since these data are mean values, not values for individuals, a regression line is not given.

about age 38 years, with 90% of peak bone mass being achieved by 18 years. Genetic and lifestyle factors influence bone mineral accrual during growth. Bone mineral density is usually measured by dual energy X-ray absorption, and the rate of bone remodeling is assessed by biomarkers (vide supra).

A proportion of ingested fluoride is retained in the body. This has been shown in fluoride balance studies where, in addition to urinary fluoride excretion, fecal fluoride was measured in infants [28] and adults [29]. At least 99% of the body burden of fluoride is associated with the skeleton [13]. Incorporation of fluoride into bone occurs in several stages: first, exchange of ions in the loosely integrated sheath surrounding the bone crystallite; second, incorporation into the hydration shell; and, finally, migration of the fluoride ion into the crystal structure during recrystallization. The first stage is rapid, occurring within 60 min of intravenous injection of fluoride. Uptake by bone is continuous, unlike in other (soft) tissues where a plateau is quickly reached [13]. In addition, the crystallites of developing bone are small in size (compared with mature bone), large in number and heavily hydrated [13]. Thus, new

bone may reflect contemporary fluoride exposure, while mature bone reflects historic fluoride exposure.

There is a steady-state relationship between the fluoride concentrations in the extracellular fluids and the hydration shells of bone crystallites [13]. According to this concept, there is a net transfer of fluoride from plasma to the hydration shells when the plasma concentration is rising and in the opposite direction when the plasma concentration is falling. In other words, the surface bone compartment is considered to be small but rapidly exchangeable when compared with the bulk, virtually non-exchangeable, inner compartment [30]. Thus, the initial uptake of ingested fluoride by bone occurs on the endosteal and periosteal surfaces which also have the highest concentrations in rats [31–33] and humans [34]. In this sense, bone surface fluoride concentrations have been shown to increase in rats in the first hours or even days following an acute high dose of fluoride [33, 35, 36], and could be useful to clarify the causa mortis when fluoride is suspected as the cause of death. However, the suitability of this biomarker to humans has not been evaluated so far. It is important to consider that bone surface fluoride concentrations are not expected to be homogeneous among humans in defined populations or regions. Bone fluoride concentrations increase with increase of past fluoride intake [37] and age [38–40]. They also seem to be influenced by acid-base balance (decrease in cases of metabolic alkalosis and increase in high altitude) and genetic background [27, 41]. Because of these variables, surface bone fluoride concentrations per se may not be the best indicators of acute exposure to lethal amounts of fluoride and the ratio surface-to-interior concentrations should be preferred.

The choice of site for bone sampling needs careful consideration, since there is much variation in fluoride concentration between sites [42]. Since cancellous bone is much more biologically active than compact bone, the highest values might be obtained in the metaphysis cancellous bone (>9,000 mg/kg in ash), while within a bone such as the rib, cancellous bone might have a fluoride concentration of 3,500 mg/kg (in ash) compared with 1,700 mg/kg (in ash) in rib compact bone [38]. The suitability of bone as a marker of historic fluoride exposure is considered further in the chapter by Pessan and Buzalaf [this vol., pp. 52–65].

Saliva

Knowledge of the fluoride concentration in saliva has been considered important, principally because it influences plaque fluoride concentration strongly and hence the control of dental caries. There are many examples of the fluoride concentration in whole saliva increasing in response to the provision of caries-prevention agents [43–48]. Because whole saliva (usually collected by drooling into a container) is contaminated with fluoride from food and therapeutic agents, saliva has been collected from parotid and submandibular/sublingual ducts, by specially constructed devices. The fluoride concentration in ductal saliva has been studied in relation to: (1) plasma fluoride concentration, (2) salivary stimulation and flow rate, and (3) fluoride ingestion.

Fluoride concentrations in whole saliva and ductal saliva were compared by Yao and Gron [49]. The concentration in whole saliva (mean of 6 subjects) was 15.8 ng/ml (0.83 µmol/l), while after centrifuging, the concentration was 9.3 ng/ml (0.49 µmol/l). The corresponding values for parotid and submandibular ductal saliva were 6.5 ng/ml (0.34 µmol/l) and 6.3 ng/ml (0.33 µmol/l), respectively. In a series of three experiments, although seemingly using the same subjects, Oliveby et al. [17–19] recorded fluoride concentrations in parotid and submandibular ductal saliva and in whole saliva (as collected) of 0.17 µmol/l (3.23 ng/ml), 0.46 µmol/l (8.7 ng/ml) and 0.71 µmol/l (13.5 ng/ml), respectively. Also in the study by Whitford et al. [20], fluoride concentrations in whole saliva

were not significantly related with plasma fluoride concentrations of 5- to 10-year-old children (n = 17) while parotid fluoride concentrations were (by a proportionality constant of 0.8). A further comparison of fluoride concentrations in parotid and submandibular ductal saliva was published by Twetman et al. [50]; the subjects were 12 young adolescent girls. Fluoride concentrations (stimulated flow) were higher in submandibular saliva (0.55 µmol/l, 10.5 ng/ml) than in parotid saliva (0.25 µmol/l, 4.8 ng/ml), and concentrations remained higher in submandibular duct saliva after ingestion of fluoride. Thus, fluoride concentrations appear to be slightly higher in submandibular saliva than in parotid saliva.

The series of studies by Oliveby et al. [17–19] and the study by Whitford et al. [20] clearly demonstrated the relationship between fluoride concentrations in plasma and parotid or submandibular ductal saliva. In agreement with previous research [51], Oliveby et al. [17, 18] reported that fluoride concentrations were lower in ductal saliva than in plasma. The ratios of saliva to plasma fluoride concentrations, under resting conditions, were 0.32 to 0.55 for parotid saliva [17] and 0.61 to 0.88 for submandibular saliva [18]. In contrast, in the same series of experiments, the ratio for fluoride concentrations in whole saliva (as collected) and plasma was 1.10 [19], suggesting that whole saliva had acquired fluoride from the oral environment. Whitford et al. [20] also found the fluoride concentration in whole saliva much higher than in parotid ductal saliva. The fluoride concentration in ductal parotid saliva was strongly correlated with plasma fluoride concentration, with a proportionality constant of 0.80 for the saliva/plasma relation. It follows that, for comparisons with plasma fluoride concentration and therefore the body burden of fluoride, it is preferable to examine ductal saliva rather than whole saliva.

After ingestion of fluoride, there is a close relationship between the rise in fluoride concentration in plasma and the rise in fluoride concentration in parotid and submandibular ductal salivas

[17, 18, 51]. In the study of Ekstrand [51], where 3 young adults received 3 mg fluoride, both plasma and salivary (parotid) fluoride concentrations peaked at about 30 min and remained elevated for 8 h. Salivary fluoride concentration followed that of plasma closely throughout the 8 h, with an average saliva/plasma ratio of 0.63. Commenting on the stability of this ratio during the experiment, the author commented 'that saliva [fluoride] may be used as a substitute for blood sampling in studies concerning the pharmacokinetics of fluoride'. Rather similar results were reported by Oliveby et al. [17, 18] and Whitford et al. [20] for parotid and submandibular saliva. While the stability of the saliva/plasma ratio was good for submandibular saliva (approximately 0.4–0.6), it was less good for parotid saliva (approximately 0.30–0.65) [17, 18]. Whitford [13] reported that the saliva to plasma ratio for fluoride concentration, after ingestion of 10 mg fluoride by adults, was higher for submandibular (about 0.88) than for parotid (about 0.78) ductal saliva. This was very close to the value of 0.80 reported by Whitford et al. [20] for children.

The times of the peak concentrations of fluoride in plasma and ductal saliva, after fluoride ingestion, have been studied. Ekstrand [51] found that both peaked at around 30 min after a 3 mg dose. In slight contrast, Oliveby et al. [17, 18] found: (1) that peaks were not reached until about 40 min after ingestion of 1 mg fluoride, and (2) a delay of about 10–15 min in the fluoride concentration peak in parotid saliva compared with plasma, although the results for submandibular saliva were variable.

The concentrations of many constituents of saliva change markedly after stimulation of salivary flow [13]. However, a number of researchers have shown that fluoride concentration is remarkably stable. Shannon et al. [52] found that the fluoride concentration in parotid saliva fell from 22 ng/ml (1.16 μmol/l) before stimulation to 17 ng/ml (0.89 μmol/l) after stimulation, reducing only slightly as the intensity of stimulation increased. Oliveby

et al. [17, 18] also found fluoride concentrations in parotid [17] and submandibular ductal saliva [18] were little affected by stimulation of salivary flow, for up to 2 h after fluoride ingestion, while Twetman et al. [50] reported slightly higher fluoride concentrations in unstimulated ductal saliva, before and after fluoride ingestion.

Thus, in conclusion, as a marker of plasma fluoride concentration, it would appear that submandibular/sublingual duct saliva is preferable to parotid duct saliva, and both are preferable to whole saliva, because (1) fluoride concentrations are higher, and (2) saliva to plasma ratios are more stable after fluoride ingestion. However, it should be noted that collection of submandibular/sublingual saliva is technically more difficult [18].

Sweat

Early work suggested that the concentration of fluoride in sweat was substantial: McClure et al. [12] reported 0.3–1.8 mg/l, Crosby and Shepherd [53] 0.3–0.9 mg/l, and Largent [54] 0.3–0.9 mg/l. However, estimates made after the introduction of the ion-specific electrode were very much lower. Even after ingesting 10 mg fluoride, which raised plasma concentration to 0.24 mg/l (12.6 μmol/l), the concentration in sweat was only about 0.05 mg/l (2.6 μmol/l) [11]. In a brief summary, Whitford [13] stated that fluoride concentrations in sweat were similar to concentrations in plasma (1–3 μmol/l; 0.019–0.057 mg/l; 19–57 ng/ml). Presently, issues of collection, including contamination, and lack of supporting data, preclude the use of sweat as a viable marker of contemporary fluoride exposure.

Human Milk

Backer Dirks et al. [55] estimated the fluoride concentration in human milk of mothers living in fluoridated or non-fluoridated communities, using both gas-liquid chromatography and the ion-specific electrode. The total fluoride concentrations in the milk of mothers in the two areas, respectively, were 52 and 46 ng/ml (2.7 and 2.4 µmol/l), while the ionic fluoride concentrations were 8 and 4 ng/ml (0.42 and 0.21 µmol/l). A more recent estimate of the fluoride concentration in milk from 57 lactating Turkish women was 19 ng/ml (1.0 µmol/l) [56]. The only direct comparison of the fluoride concentrations in human plasma and milk was published by Ekstrand et al. [16]. Five mothers were given 1.5 mg fluoride after fasting, and fluoride concentrations in plasma and milk were followed for 2 h. While the plasma fluoride concentration rose from about 0.6 µmol/l (11 ng/ml) to a peak of about 5 µmol/l (95 ng/ml) after 30 min, there was virtually no change in the fluoride concentration in milk which remained at about 0.26 µmol/l (4.9 ng/ml). In conclusion, on the basis of these limited data, human milk would seem unsuitable for estimating the body burden of fluoride.

Urine

Fluoride concentrations in body fluids (e.g. urine, plasma, serum, saliva) are generally recognized as being the most suitable for evaluating short-term fluoride exposures or fluoride balance (intake minus excretion), while some earlier sources suggest that samples obtained from fasting persons may be useful for estimating chronic fluoride intake or potential bone fluoride concentrations [57, 58]. Examples of the association between estimated fluoride intakes (or mass-normalized intakes) and measured fluoride concentrations in urine, plasma, and serum for individuals and groups were shown in a recent review [table 2.16; 59].

In order to be considered a viable fluoride biomarker, a relationship must be established between some property (mass, concentration, others) of the candidate biomarker and fluoride exposure or intake over a period of time: thus, from an assessment of that property, fluoride exposure might be reliably inferred. Another important point that has to be considered is whether the fluoride biomarker can be used on an individual or group (e.g. community) basis (vide infra).

Based on pharmacokinetic findings, urinary fluoride is considered a contemporary biomarker of fluoride exposure, since varying proportions of a given fluoride dose are completely excreted with the urine in less than 24 h in children and adults [13, 15]. It is important that the selected biomarker is clearly related to the fluoride exposure. Regarding urine as a possible biomarker, early studies [44, 60, 61] attempted to establish such a relationship by comparing urinary fluoride concentration and fluoride exposure. In some studies, exposure was estimated semiquantitatively, e.g. reporting the fluoride concentration of the drinking water that the study subjects ingested [44, 60, 61]. In these studies, no simple numerical relationship could be established between fluoride exposure and urinary fluoride concentration, and thus they are of limited use. Other studies have measured experimentally the amount of fluoride ingested from drinking water, foods, other beverages and, eventually, other fluoride sources such as dental hygiene products [29, 62–65]. As discussed by Villa et al. [66] a 24-hour urinary collection is the minimal recommended period of time in order to obtain good estimations of the daily amount of fluoride excretion. The daily urinary fluoride excretion is the variable generally recommended for the estimation of the daily fluoride exposure. The amount of excreted fluoride is easily obtained multiplying the 24-hour urinary volume by its fluoride concentration [62, 67].

Factors Affecting Urinary Fluoride Excretion
Changes in chronic exposure to fluoride will tend to alter plasma and bone fluoride concentrations (as discussed in previous sections of this chapter), and a number of factors can modify

Table 2. Observed range of values for 24-hour urinary fluoride excretion and estimated 24-hour fluoride exposure in young children and adults according to different fluoridation conditions

Range of urinary fluoride excretion found, mg/24 h	Range of predicted fluoride intake, mg/24 h	Fluoridation condition
Young children (≤6 years)		
0.17–0.31	0.40–0.80	low-fluoridated areas[1]
0.31–0.50	0.80–1.34	optimally fluoridated areas[2]
>0.60	>1.63	higher than 'optimal'[3]
Adults (18 to 50+ years)		
1.00–1.40	1.31–2.05	low-fluoridated areas[1]
1.50–2.50	2.24–4.10	optimally fluoridated areas[2]
>3.00	>5.00	higher than 'optimal'[4]

Data taken from Villa et al. [66], with permission (n = 212 young children and 269 adults).
[1] Fluoride concentrations in drinking water ≤0.4 mg/l.
[2] Fluoride concentrations in drinking water in the range 0.6–1.0 mg/l.
[3] On a chronic basis, fluoride exposure might cause an objectionable prevalence of moderate and severe enamel fluorosis.
[4] On a chronic basis, fluoride exposure might cause a preclinical stage of skeletal fluorosis [59].

fluoride pharmacokinetics, providing another way to change fluoride tissue concentrations [13]. Fluoride clearance tends to increase with urinary pH [68]. One proposed mechanism for this is decreased reabsorption in the renal tubule, since hydrogen fluoride (which is formed at lower pH values) easily crosses cell membranes, which are nearly impermeable to the fluoride ion [68]. Thus, increasing urinary pH tends to reduce fluoride retention [68]. As a result, fluoride retention might be affected by environments or conditions that chronically affect urinary pH, including diet, drugs, altitude and certain diseases, e.g. chronic obstructive pulmonary disease [13].

Since the kidney is the major route of fluoride excretion, it is not surprising that increased plasma and bone fluoride concentrations have been observed in patients with kidney disease. Plasma fluoride concentrations have also been demonstrated to be elevated in patients with severely compromised kidney function, where glomerular filtration rates are reduced to around 20% of normal, as measured via creatinine clearance or serum creatinine concentrations [58, 69, 70].

Urine as a Contemporary Fluoride Biomarker
While the first reports of simultaneous measurement of total daily fluoride intake (TDFI) and daily urinary fluoride excretion (DUFE) were published many years ago [29, 71, 72], several studies have been undertaken more recently in children [28, 62, 63, 65, 73, 74] and adults [64, 67]. A recent paper [66] reviewed all of the available published data on the simultaneous experimental assessment of TDFI and DUFE using individuals' data from each study, and the relationship between these variables was examined in order to assess the suitability of DUFE as a predictor (biomarker) of TDFI.

Results obtained from nine independent studies (n = 212) carried out in young children (0.15–7 years old) from six different Western countries,

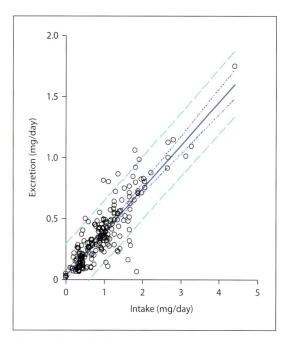

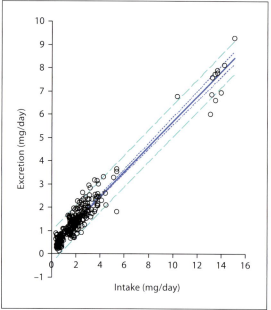

Fig. 2. Relationship between daily urinary fluoride excretion and total daily fluoride intake for 212 young children aged 0.15–7 years recorded in 9 studies in 6 countries. The full line is the best fit; the inner interrupted lines indicate the 95% CI of the regression, and the outer interrupted lines indicate the 95% prediction interval. Reproduced from Villa et al. [66], with permission.

Fig. 3. Relationship between daily urinary fluoride excretion and total daily fluoride intake for 269 data pairs from adults aged 18–75 years recorded in 8 studies in 2 countries. The full line is the best fit; the inner interrupted lines indicate the 95% CI of the regression, and the outer interrupted lines indicate the 95% prediction interval. Reproduced from Villa et al. [66], with permission.

and six independent studies (n = 269) in adults (18- to 75-years-old) from two American countries, were available for the analysis (fig. 2 and 3). Highly significant linear relationships were found between DUFE and TDFI for both children and adults. The values of the intercepts and slopes of the best fit regressions for both age groups were significantly different [66], indicating that the proportion of fluoride retained in children's hard tissues is higher than the corresponding values for adults, in agreement with previous knowledge. No differences due to gender were observed for both age groups. These results strongly suggest that the daily urinary fluoride excretion is a reasonably good biomarker of contemporary fluoride exposure.

Individual or Group Biomarker?

Statistical analysis of the above-mentioned linear relationships clearly show that the 95% prediction intervals associated with the regression lines do not suggest that DUFE is viable as a precise estimator of TDFI on an individual basis. However, the 95% CI bands do suggest DUFE is appropriate when considering fluoride exposure on a group (or community) basis [66]. Thus, it can be concluded that, at this time, urinary fluoride excretion has a very limited value as a biomarker of individual fluoride exposure. This situation is similar to that of other fluoride biomarkers discussed in this chapter and that of Pessan and Buzalaf [this vol., pp. 52–65]. From an epidemiological point of view, estimating the average (and 95% CI) daily fluoride exposure on a community basis might have some merit. When a 24-hour urinary fluoride excretion study is performed on a relatively high (n ≥20 according to general practice in most published reports on this type of study) number of subjects, and accepting that the distribution of frequencies of urinary excretion values is nearly normal, the average and 95% CI values might be considered reliable estimators of 24-hour fluoride exposure for the group or community from which the sample was taken. This is especially valid when such a group or community of individuals is under 'customary' fluoride intake conditions, i.e. when the different fluoride sources remain stable over time. It might be argued that 24-hour urine collection provides a 'snap-shot' of the fluoride exposure for the particular study day, since each individual might present a value that is different from the one he/she would have presented the following day or the day before. However, using a statistical approach, it can be considered that most of the within- and inter-individual variation over time would be virtually cancelled out when average values are considered. Under the above-mentioned conditions, several urinary fluoride excretion studies have been considered useful for evaluating the safety of fluoride-based systemic preventive caries programs [75]. Alternatively, an indication of a higher than safe fluoride exposure (in terms of the risk of an undesirable prevalence of enamel fluorosis in children, or skeletal fluorosis in adults, as defined in table 2) can be also established from community studies of urinary fluoride excretion, provided that updated guidelines on 'expected' average value ranges are available for different fluoride exposure conditions. Several health conditions that cause fluoride renal clearance disturbances have been mentioned previously, but a detailed discussion of particularly vulnerable subgroups of subjects is beyond the scope of this section. For a thorough discussion of this subject, see a recently published review by the US Environmental Protection Agency [chapter 3; 59]. In addition, it is important to take into account that, among healthy subjects, the effect of diet on their urinary pH might have a significant effect on observed urinary fluoride excretion values for a certain range of average fluoride exposures [13].

Fluoride Urinary Excretion: Normal or Observed Range of Values?

Fluoride is not homeostatically controlled [13]. This means that fluoride levels in body fluids or tissues will show a range of values that will be highly variable and dependent on short- and long-term fluoride exposure. Thus, a 'normal' (as used in clinical chemistry) urinary fluoride excretion range of values cannot be established. In the introductory section of this chapter, the WHO statement on the usefulness of a fluoride biomarker said that 'a fluoride biomarker is of value primarily for identifying and monitoring deficient or excessive intakes of biologically available fluoride' [1]. In this context, a semi-quantitative use of fluoride urinary excretion as a biomarker was presented in an earlier monograph [75]. The latter publication presents ranges of provisional standard values for daily fluoride urinary excretion in young children in low and 'optimally' fluoridated areas [table 5; 75].

Based on the available data, a quantitative estimation of fluoride exposure can now be proposed using fluoride urinary excretion values and the numerical relationships between DUFE and TDFI [66]. Thus, for sets of observed ranges of values of 24-hour urinary fluoride excretion, the estimated daily fluoride exposure in young children and adults under different fluoridation conditions can be obtained (table 2).

The predicted values for 24 h fluoride exposures in young children and adults (table 2) were estimated using published numerical models [66] and are presented here as a provisional guideline that shows the feasibility of using daily fluoride urinary excretion as a biomarker of daily fluoride exposure. However, certain limitations apply: first, the numerical values of predicted fluoride exposures appearing in table 2 have estimated 95% CI of 10–15%; second, the available studies from which the experimental data were taken included subjects who consumed 'westernized' diets [66]. Thus, for those communities where 'non-westernized' more vegetarian diets are consumed, it is reasonable to assume that urinary pH values could be higher [13] and, consequently, the proportion of fluoride excreted with the urine might be higher. Further studies on this latter issue are needed. It is conceptually reasonable to point out that different age subgroups of the two age groups (young children and adults) from which the present estimates were obtained, might present different numerical values for the relationship between DUFE and TDFI, especially so among young children when the different rates of bone formation of infants and 6-year-olds are considered, as discussed in previous reports [13, 66, 76]. However, the imprecision associated with the variability caused by inter-individual physiological differences might obscure these effects.

Conclusion

Interest in the identification of viable and accurate biomarkers for fluoride exposure has increased significantly over the past several decades. This is partly due to the increasing recognition that the therapeutic ratio for fluoride is relatively low and the need to avoid disbenefits, partly because of technical advances in measurement, and partly because of the expansion in knowledge of fluoride metabolism. Fluoride toxicity can be acute or chronic, and biomarkers are needed for assessing both situations. Contemporary biomarkers assess present (acute) exposure and might include fluoride concentration in blood, bone surface, saliva, milk, sweat and urine, while past exposure is assessed by historic biomarkers, which might include bone, teeth, nail and hair. Contemporary biomarkers have been discussed in this chapter.

There have been a number of reports of fluoride concentration in plasma, both after fasting and after ingestion of known doses of fluoride. Maximum concentration occurs at about 30 min and resting values may not be reached until 3–6 h after exposure. Apart from fluoride dose, plasma fluoride concentration varies with site of blood collection, age, acid-base balance, altitude, hematocrit and genetic background. At present, there are insufficient data on plasma fluoride concentration, for various age groups, to determine 'normal' plasma fluoride concentrations.

Much of the unexcreted fluoride that is eventually incorporated into bone is found within the bone crystal structure. Although there has been some attention focused on bone surface as a potential marker of contemporary fluoride exposure and, in particular, the use of the ratio of surface-to-interior concentrations, there are insufficient data for recommending the use of bone as a viable marker for estimating contemporary fluoride exposure in humans.

Fluoride concentrations in parotid and submandibular/sublingual ductal saliva follow plasma fluoride concentration, although at a lower concentration. Submandibular saliva may have advantages over parotid saliva, although its collection is more difficult. At present, there are insufficient data to establish a normal range of fluoride concentrations in ductal saliva as a basis for recommending saliva as a marker of fluoride exposure. Sweat and human milk are unsuitable as biomarkers of fluoride exposure.

A proportion of ingested fluoride is excreted in urine. This proportion is influenced by age, urinary pH, and the several conditions that affect urinary pH. Plots of daily urinary fluoride excretion against total daily fluoride intake, for young children and adults separately, reveal different slopes of the regression lines, indicating the relationship is different in the two age groups. The plots also suggest that daily urinary fluoride excretion is suitable for predicting fluoride intake for groups of people, but not for individuals.

While fluoride concentrations in plasma, saliva and urine have some ability to predict fluoride exposure, data are, at present, insufficient to recommend fluoride concentration in these body fluids as viable biomarkers of contemporary fluoride exposure for individuals. Fluoride concentration in urine can be considered a useful biomarker of contemporary fluoride exposure for groups of people, and normal values have been published.

Future research on biomarkers for contemporary fluoride exposure should focus on urine, saliva, plasma and bone, listed in order of importance. Areas of uncertainty have been indicated within the relevant sub-sections.

References

1 WHO: Fluorides and oral health. Technical Report Series 846. Geneva, WHO, 1994.
2 Selwitz RH: Introduction to the workshop on methods for assessing fluoride accumulation and effects in the body. Adv Dent Res 1994;8:3–4.
3 Medical Research Council: Water fluoridation and health (working group report). London, Medical Research Council, 2002.
4 Trull AK, Demers LM, Holt DW, Johnston A, Tredger JM, Price CP: Biomarkers of Disease: An Evidence-Based Approach. Cambridge, Cambridge University Press, 2002.
5 Risteli J, Kauppila S, Jukkola A, Marjoniemi E, Melkko J, Risteli L: Biomarkers of bone formation; in Trull AK, Demers LM, Holt DW, Johnston A, Tredger JM, Price CP (eds): Biomarkers of Disease: An Evidence-Based Approach. Cambridge, Cambridge University Press, 2002, pp 115–121.
6 Eastell R, Hart S: Bone turnover markers in clinical practice; in Trull AK, Demers LM, Holt DW, Johnston A, Tredger JM, Price CP (eds): Biomarkers of Disease: An Evidence-Based Approach. Cambridge, Cambridge University Press, 2002, pp 99–114.
7 Robins SP: Biochemical markers of bone resorption; in Trull AK, Demers LM, Holt DW, Johnston A, Tredger JM, Price CP (eds): Biomarkers of Disease: An Evidence-Based Approach. Cambridge, Cambridge University Press, 2002, pp 122–132.

8 Jones R: Using intelligent systems in clinical decision support; in Trull AK, Demers LM, Holt DW, Johnston A, Tredger JM, Price CP (eds): Biomarkers of Disease: An Evidence-Based Approach. Cambridge, Cambridge University Press, 2002, pp 32–44.
9 Marshall WJ, Bangert SK: Clinical Chemistry, ed 6. Edinburgh, Mosby, 2008, pp 1–13.
10 Armstrong WD, Gedalia I, Singer L, Weatherell JA, Weidmann SM: Distribution of fluorides; in WHO (ed): Fluorides and Human Health. Geneva, WHO, 1970, pp 93–139.
11 Murray JJ, Rugg-Gunn AJ, Jenkins GN: Fluorides in caries prevention, ed 3. Oxford, Butterworth-Heinemann, 1991, pp 262–294.
12 McClure FJ, Mitchell HH, Hamilton TS, Kinser CA: Balances of fluorine ingested from various sources in food and water by five young men. J Ind Hyg Toxicol 1945;27:159–170.
13 Whitford GM: The Metabolism and Toxicology of Fluoride, ed 2. Basel, Karger, 1996.
14 Ekstrand J, Alvan G, Boreus LO, Norlin A: Pharmacokinetics of fluoride in man after single and multiple oral doses. Europ J Clin Pharmacol 1977;12:311–317.
15 Ekstrand J: Relationship between fluoride in the drinking water and the plasma fluoride concentration in man. Caries Res 1978;12:123–127.
16 Ekstrand J, Boreus LO, de Chateau P: No evidence of transfer of fluoride from plasma to breast milk. BMJ 1981;283:761–762.

17 Oliveby A, Lagerlof F, Ekstrand J, Dawes C: Influence of flow rate, pH and plasma fluoride concentrations on fluoride concentration in human parotid saliva. Arch Oral Biol 1989;34:191–194.
18 Oliveby A, Lagerlof F, Ekstrand J, Dawes C: Studies on fluoride concentrations in human submandibular/sublingual saliva and their relation to flow rate and plasma fluoride levels. J Dent Res 1989;68:146–149.
19 Oliveby A, Lagerlof F, Ekstrand J, Dawes C: Studies on fluoride excretion in human whole saliva and its relation to flow rate and plasma fluoride levels. Caries Res 1989;23:243–246.
20 Whitford GM, Thomas JE, Adair SM: Fluoride in whole saliva, parotid ductal saliva and plasma in children. Arch Oral Biol 1999;44:785–788.
21 Levy FM, Bastos JR, Buzalaf MA: Nails as biomarkers of fluoride in children of fluoridated communities. J Dent Child 2004;71:121–125.
22 Maguire A, Zohouri FV, Mathers JC, Steen IN, Hindmarsh PN, Moynihan PJ: Bioavailability of fluoride in drinking water: a human experiment. J Dent Res 2005;84:989–993.
23 Cardoso VE, Whitford GM, Buzalaf MA: Relationship between daily fluoride intake from diet and the use of dentifrice and human plasma fluoride concentrations. Arch Oral Biol 2006;51:552–557.
24 Whitford GM, Sampaio FC, Pinto CS, Maria AG, Cardoso VES, Buzalaf MAR: Pharmacokinetics of ingested fluoride: lack of effect of chemical compound. Arch Oral Biol 2008;53:1037–1041.

25 Cardoso VES, Whitford GM, Aoyama H, Buzalaf MAR: Daily variations in human plasma fluoride concentrations. J Fluorine Chem 2008;129:1193–1198.

26 Buzalaf MAR, Leite AL, Carvalho NTA, Rodrigues MHC, Takamori ER, Nicolielo DB, Levy FM, Cardoso VES: Bioavailability of fluoride administered as sodium fluoride or monofluorophosphate to humans. J Fluoride Chem 2008;129:691–694.

27 Carvalho JG, Leite AL, Yan D, Everett ET, Whitford GM, Buzalaf MAR: Influence of genetic background on fluoride metabolism in mice. J Dent Res 2009;88:1054–1058.

28 Ekstrand J, Zeigler EE, Nelson SE, Fomon SJ: Absorption and retention of dietary and supplemental fluoride by infants. Adv Dent Res 1994;8:175–180.

29 Spencer H, Lewin I, Wistrowski E, Samachson J: Fluoride metabolism in man. Am J Med 1970;49:807–813.

30 Rao HV, Beliles RP, Whitford GM, Turner CH: A physiologically based pharmacokinetic model for fluoride uptake by bone. Reg Toxicol Pharmacol 1995;22:30–42.

31 Narita N, Kato K, Nakagaki H, Ohno N, Kameyama Y, Weatherell JA: Distribution of fluoride's concentration in rat's bone. Calcif Tiss Int 1990;46:200–204.

32 Li J, Nakagaki H, Kato K, Ishiguro K, Ohno N, Kameyama Y, Mukai M, Ikeda M, Weatherell J, Robinson C: Distribution profiles of fluoride in three different kinds of rats bone. Bone 1993;14: 835–842.

33 Buzalaf MAR, Caroselli EE, Carvalho JG, Oliveira RC, Cardoso VES, Whitford GM: Bone surface and whole bone as biomarkers of acute fluoride exposure. J Analyt Toxicol 2005;29:810–813.

34 Ishiguro K, Nakagaki H, Tsuboi S, Narita N, Kato K, Li J, Kamei H, Yoshioka I, Miyauchi K, Hosoe H, et al: Distribution of fluoride in cortical bone of human rib. Calcif Tiss Int 1993;52:278–282.

35 Bezerra de Menezes LM, Volpato MC, Rosalen PL, Cury JA: Bone as a biomarker of acute fluoride toxicity. Forensic Sci Int 2003;137:209–214.

36 Buzalaf MAR, Caroselli EE, Oliveira RC, Granjeiro JM, Whitford GM: Nail and bone surface as biomarkers of acute fluoride exposure in rats. J Analyt Toxicol 2004;28:249–252.

37 Charen J, Taves DR, Stamm JW, Parkins FM: Bone fluoride concentration associated with fluoridated drinking water. Calcif Tissue Int 1979;27:95–99.

38 Weidmann SM, Weatherell JA: Distribution in hard tissues; in WHO (ed): Fluorides and Human Health. Geneva, WHO, 1970, pp 104–128.

39 Parkins FM, Tinanoff N, Moutinho M, Anstey MB: Relationships of human plasma fluoride and bone fluoride to age. Calcif Tiss Res 1974;16:335–338.

40 Richards A, Mosekilde L, Sogaard CH: Normal age-related changes in fluoride content of vertebral trabecular bone – relation to bone quality. Bone 1994; 15:21–26.

41 Everett ET, McHenry MA, Reynolds N, Eggertsson H, Sullivan J, Kantmann C, Martinez-Mier EA, Warrick JM, Stookey GK: Dental fluorosis: variability among different inbred mouse strains. J Dent Res 2002;81:794–798.

42 Zipkin I, McClure FJ, Leone NC, Lee WA: Fluoride deposition in human bones after prolonged ingestion of fluoride in drinking water. Publ Health Rep 1958;73:732–740.

43 Brunn C, Givskov H, Thylstrup A: Whole saliva fluoride after toothbrushing with NaF and MFP dentrifices with different F concentrations. Caries Res 1984;18: 282–288.

44 Schamschula RG, Sugar E, Un PSH, Toth K, Barmes DE, Adkins BL: Physiological indicators of fluoride exposure and utilization: an epidemiological study. Community Dent Oral Epidemiol 1985;13: 104–107.

45 Boros I, Keszler P, Banoczy J: Fluoride concentrations of unstimulated whole and labial gland saliva in young adults after fluoride intake with milk. Caries Res 2001;35:167–172.

46 Petersson LG, Arvidsson I, Lynch E, Engstrom K, Twetman S: Fluoride concentrations in saliva and dental plaque in young children after intake of fluoridated milk. Caries Res 2002;36:40–43.

47 Hedman J, Sjoman R, Sjostrom I, Twetman S: Fluoride concentration in saliva after consumption of a dinner meal prepared with fluoridated salt. Caries Res 2006;40:158–162.

48 Pessan JP, Silva SM, Lauris JR, Sampaio FC, Whitford GM, Buzalaf MA: Fluoride uptake by plaque from water and from dentifrice. J Dent Res 2008;87: 461–465.

49 Yao K, Gron P: Fluoride concentrations in duct saliva and in whole saliva. Caries Res 1970;4:321–331.

50 Twetman S, Nederfors, Petersson LG: Fluoride concentration in whole saliva and separate gland secretions in schoolchildren after intake of fluoridated milk. Caries Res 1998;32:412–416.

51 Ekstrand J: Fluoride concentrations in saliva after single oral doses and their relation to plasma fluoride. Scand J Dent Res 1977;85:16–17.

52 Shannon IL, Suddick RP, Edmonds EJ: Effect of rate of gland function on parotid saliva fluoride concentration in the human. Caries Res 1973;7:1–10.

53 Crosby ND, Shepherd PA: Studies on patterns of fluid intake, water balance and fluoride. Med J Aust 1957;ii: 341–346.

54 Hodge HC, Smith FA, Gedalia I: Excretion of fluorides; in WHO (ed): Fluorides and human health. Geneva, WHO, 1970, pp 141–161.

55 Backer Dirks O, Jongeling-Eijndhoven JMPA, Flissebaalje TD, Gedalia I: Total and free ionic fluoride in human and cow's milk as determined by gas-liquid chromatography and the fluoride electrode. Caries Res 1974;8:181–186.

56 Koparal E, Ertugrul F, Oztekin K: Fluoride levels in breast milk and infant foods. J Clin Pediatr Dent 2000;24: 299–302.

57 Ericsson Y, Gydell K, Hammarskioeld T: Blood plasma fluoride: an indicator of skeletal fluoride content. J Int Res Commun Syst 1973;1:33–35.

58 Waterhouse C, Taves D, Munzer A: Serum inorganic fluoride: changes related to previous fluoride intake, renal function and bone resorption. Clin Sci 1980;58:145–152.

59 EPA: Fluoride in drinking water; a scientific review of EPA's standards (2006). Committee on Fluoride in Drinking Water. Board on Environmental Studies and Toxicology. Division on Earth and Life Studies. National Research Council of the National Academies. Washington, National Academies Press Washington, 2006.

60 Czarnowski W, Krechniak J: Fluoride in the urine, hair, and nails of phosphate fertiliser workers. Br J Ind Med 1990;47:349–351.

61 Czarnowski W, Krechniak J, Urbanska B, Stolarska K, Taraszewska-Czarnowska M, Muraszko-Klaudel A: The impact of water-borne fluoride on bone density. Fluoride 1999;32:91–95.

62 Villa AE, Anabalón M, Cabezas L: The fractional urinary fluoride excretion in young children under stable fluoride intake conditions. Community Dent Oral Epidemiol 2000;28:344–355.

63 Haftenberger M, Viergutz G, Neumeister V, Hetzer G: Total fluoride intake and urinary excretion in German children aged 3–6 years. Caries Res 2001;35:451–457.

64 Villa AE, Cabezas L, Anabalón M, Garza E: The fractional urinary fluoride excretion of adolescents and adults under customary fluoride intake conditions, in a community with 0.6 mg F/L in its drinking water. Community Dent Health 2004;21:11–18.

65 Maguire A, Zohouri FV, Hindmarch PN, Hatts J, Moynihan PJ: Fluoride intake and urinary excretion in 6- to 7-year-old children living in optimally, sub-optimally and non-fluoridated areas. Community Dent Oral Epidemiol 2007; 35:479–488.

66 Villa A, Anabalon M, Zohouri V, Maguire A, Franco AM, Rugg-Gunn A: Relationships between fluoride intake, urinary fluoride excretion and fluoride retention in children and adults: an analysis of available data. Caries Res 2010;44:60–68.

67 Villa AE, Anabalón M, Rugg-Gunn A: Fractional urinary fluoride excretion of young female adults during the diurnal and nocturnal periods. Caries Res 2008; 42:275–281.

68 Whitford GM, Pashley DH, Stringer GI: Fluoride renal clearance: a pH-dependent event. Am J Physiol 1976;230:527–532.

69 Schiffl HH, Binswanger U: Human urinary fluoride excretion as influenced by renal functional impairment. Nephron 1980;26:69–72.

70 Hanhijärvi H, Penttilä I: The relationship between human ionic plasma fluoride and serum creatinine concentrations in cases of renal and cardiac insufficiency in a fluoridated community. Proc Finn Dent Soc 1981;77:330–335.

71 Ekstrand J, Hardell LI, Spak CJ: Fluoride balance studies on infants in a 1 ppm water fluoride area. Caries Res 1984;18:87–92.

72 Maheshwari UR, McDonald JT, Schneider VS, Brunetti AJ, Leybin L, Newbrun E, Hodge HC: Fluoride balance studies in ambulatory healthy men with and without fluoride supplements. Am J Clin Nutr 1981;34:2679–2684.

73 Zohouri FV, Rugg-Gunn AJ: Total fluoride intake and urinary excretion in 4-year-old Iranian children residing in low-fluoride areas. Br J Nutr 2000;83:15–25.

74 Franco AM, Saldarriaga A, Martignon S, González MC, Villa AE: Fluoride intake and fractional urinary fluoride excretion of Colombian preschool children. Community Dent Health 2005;22:272–278.

75 Marthaler TM: Monitoring of renal fluoride excretion in community preventive programmes on oral health. Geneva, World Health Organization, 1999.

76 Whitford G: Fluoride metabolism and excretion in children. J Public Health Dent 1999;59:224–228.

Prof. Andrew John Rugg-Gunn
Morven, Boughmore Road
Sidmouth
Devon EX10 8SH (UK)
Tel. +44 1395 578746, E-Mail andrew@rugg-gunn.net

Buzalaf MAR (ed): Fluoride and the Oral Environment.
Monogr Oral Sci. Basel, Karger, 2011, vol 22, pp 52–65

Historical and Recent Biological Markers of Exposure to Fluoride

Juliano Pelim Pessan[a] · Marília Afonso Rabelo Buzalaf[b]

[a]Department of Pediatric Dentistry and Public Health, Araçatuba Dental School, São Paulo State University, Araçatuba, and
[b]Department of Biological Sciences, Bauru Dental School, University of São Paulo, Bauru, Brazil

Abstract

Recent and historical biomarkers assess chronic or sub-chronic exposure to fluoride. The most studied recent biomarkers are nails and hair. Both can be non-invasively obtained, although collection of nails is more accepted by the subjects. External contamination may be a problem for both biomarkers and still needs to be better evaluated. Nails have been more extensively studied. Although the available knowledge does not allow their use as predictors of dental fluorosis by individual subjects, since reference values of fluoride have not yet been established, they have a strong potential for use in epidemiological surveys. Toenails should be preferred instead of fingernails, and variables that are known to affect nail fluoride concentrations – such as age, gender and geographical area –should be considered. The main historical biomarkers that could indicate total fluoride body burden are bone and dentin. Of these, bone is more studied, but its fluoride concentrations vary according to the type of bone and subjects' age and gender. They are also influenced by genetic background, renal function and remodeling rate, variables that complicate the establishment of a normal range of fluoride levels in bone that could indicate 'desirable' exposure to fluoride. The main issue when attempting to use bone as biomarker of fluoride exposure is the difficulty and invasiveness of sample collection. In this aspect, collection of dentin, especially from 3rd molars that are commonly extracted, is advantageous. However, mean values also span a wide range and reference concentrations have not been published yet.

Measuring fluoride intake is an important tool when controlling the risk factors for dental fluorosis. The determination of ingestion levels of fluoride, however, is becoming increasingly difficult, as fluoride comes from different sources. Considering that only the absorbed fluoride is implicated in the development of dental fluorosis, the monitoring of fluoride absorption, instead of fluoride intake, seems to be more accurate [1].

Monitoring fluoride exposures can be accomplished through the analysis of several biological tissues or fluids, with varying degrees of accuracy. While studies evaluating the time course of body fluid fluoride concentrations following the ingestion of a fluoride compound typically involve timed collections of blood plasma or parotid ductal saliva, the analysis of urinary fluoride concentrations can also provide useful information, although the data are less precise and more difficult to interpret than those derived from the analysis of plasma [2]. In order to have more precise indicators of the fluoride levels in the organism and, therefore, to be able to predict the risk of dental fluorosis, the search for biomarkers of exposure to fluoride has been intensified over the last years. Biological markers or 'biomarkers' are defined as indicators signaling events in biological systems or samples [3]. According to Grandjean [4], biomarkers are not used as diagnostic tests, but instead as indicators of changes that could lead to a clinical disease.

Biomarkers of fluoride can be arranged according to two classifications. The Committee of Biomarkers of the National Research Council [3] divided biomarkers into biomarkers of: (1) effect, (2) susceptibility, and (3) exposure; while the World Health Organization [5] proposes a time-perspective classification, in which biomarkers are divided into: (1) historical, (2) contemporary, and (3) recent markers. The present review will adopt the WHO criteria and will focus on recent (nails and hair) and historical (bone and teeth) biological markers of exposure to fluoride. Contemporary biomarkers are addressed by Rugg-Gunn et al. [this vol., pp. 37–51].

Recent Biomarkers of Exposure to Fluoride

Nails

Among the short- and long-term biomarkers studied, nails seem to be promising both for acute [6], subchronic [7–9] and chronic [7, 10–15] exposure to fluoride. Nail sampling has some advantages, since samples can be accessed and collected in a non-invasive manner, besides the possibility of storage for long periods of time without degradation. Also, the concentration of fluoride reflects the average level of intake and plasma concentration over a protracted period, in contrast to the analysis of urine, plasma or ductal saliva, whose fluoride concentrations are more like 'snapshots', and therefore subject to change due to recent fluoride intake and certain physiological variables. Finally, fingernail concentrations are not affected by variables such as fluoride intake within the last few hours or differences in glomerular filtration rate, urinary pH or urinary flow rate. Such advantages make the analysis of fingernail clippings an attractive alternative to other body fluids or tissues for the purpose of monitoring fluoride exposure [2].

The idea of using nails to monitor fluoride exposure is not new. Studies conducted in the 1970s and 1980s reported that fluoride concentrations in nails could reflect differences in chronic fluoride exposure from the atmosphere [16] and from water [17]. The possibility of using nails as biomarkers of subchronic exposure to fluoride was first described by Whitford et al. [7], about 10 years ago. In that study, one of the authors increased his fluoride intake for 1 month (3–6 mg/day) and this was reflected in fingernail fluoride concentrations 3.5 months later (fig. 1). That study also helped to clarify how fluoride is incorporated in nails. Although incorporation of fluoride from the environment is also possible, it became clear that fluoride enters the nail mainly via the growth end, and that the concentration in the nail clipping is determined by the average plasma fluoride concentration that existed while the clipping was forming [2]. Incorporation of fluoride through the nail bed seems to contribute at a lesser extent to the total nail fluoride concentration. In a study conducted in 2- to 3-year-old children, a fluoride-free toothpaste was used for 1 month, then a fluoridated toothpaste (1,570 ppm as sodium monofluorophosphate) for another month, and finally a fluoride-free toothpaste for an extra month [8]. Although the highest peak fluoride concentration was seen 16 weeks after starting use of the fluoridated toothpaste, a smaller peak was also seen 12 weeks earlier, which could suggest incorporation of fluoride also through the nail bed, although to a lesser extent (fig. 2). In a subsequent study, using the same protocol as that described by Whitford et al. [7], subjects increased their fluoride intake at a much lower level (1.8 mg/day), and nail growth rates and lengths were also evaluated. The increases verified in nails occurred within the limits of the

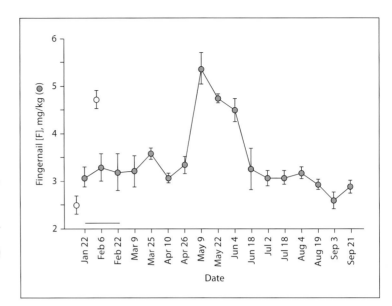

Fig. 1. Fingernails fluoride concentrations over time after 1 subject increased his fluoride intake by 3 mg/day for 30 days (horizontal bar). Open circles indicate mean urinary fluoride concentrations before (left) and after (right) increased fluoride intake.
Source: Whitford et al. [7].

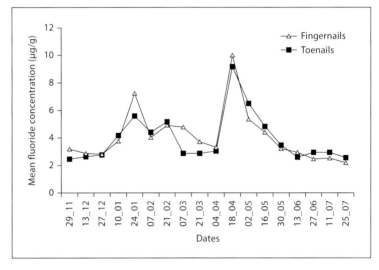

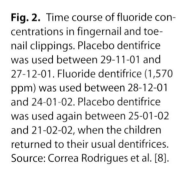

Fig. 2. Time course of fluoride concentrations in fingernail and toenail clippings. Placebo dentifrice was used between 29-11-01 and 27-12-01. Fluoride dentifrice (1,570 ppm) was used between 28-12-01 and 24-01-02. Placebo dentifrice was used again between 25-01-02 and 21-02-02, when the children returned to their usual dentifrices.
Source: Correa Rodrigues et al. [8].

95% CI of the mean lag times for fluoride detection in nails (fig. 3, 4), indicating that nail growth rates and lengths are important determinants when evaluating subchronic fluoride exposure [9].

Table 1 summarizes the main published studies on the relationship between fluoride exposure and its concentrations in nails. According to the table, fingernails were found to have higher F concentrations than toenails in 3 of the 7 studies that compared fluoride concentrations between the two types of nails [7, 12, 14]. The authors attribute the higher F concentrations in fingernails to

Fig. 3. Time course of fluoride concentrations in fingernails clippings (n = 10). The clippings were obtained every 14 days, totalizing 15 fingernails clippings. An additional 1.8 mg/day of fluoride was ingested in 3 divided doses for 30 days from the beginning of the study (horizontal bar). Fluoride intake from the diet and dentifrice was estimated in 1.5 mg/day throughout the 6-month period. After 98, 112 and 154 days the additional intake of 1.8 mg/day was reflected in increased fluoride concentrations in toenails, although this difference was not statistically significant from baseline values. After day 154, fluoride concentrations decreased, and the values obtained in days 182 and 210 were significantly lower. Data are presented as mean ± SE. Different letters indicate statistical significance (p < 0.05). Vertical bars indicate expected mean lag time for fluoride detection in fingernails (solid) and 95% confidence interval (dots). Source: Buzalaf et al. [9].

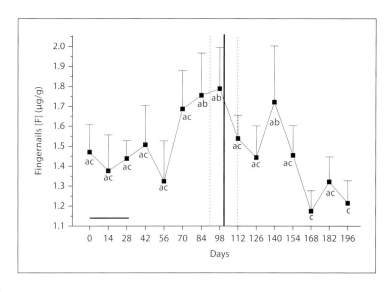

Fig. 4. Time course of fluoride concentrations in toenails clippings (n = 10). The clippings were obtained every 14 days, totalizing 15 toenails clippings. An additional 1.8 mg/day of fluoride was ingested in 3 divided doses for 30 days from the beginning of the study (horizontal bar). Fluoride intake from the diet and dentifrice was estimated in 1.5 mg/day throughout the 6-month period. After 126 and 154 days the additional intake of 1.8 mg/day was reflected in significantly increased fluoride concentrations in toenails. Data are presented as mean ± SE. Different letters indicate statistical significance (p < 0.05). Vertical bars indicate expected mean lag time for fluoride detection in toenails (solid) and 95% confidence interval (dots). Source: Buzalaf et al. [9].

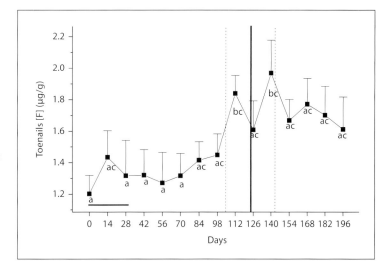

Table 1. Summary of the main studies published (in reverse chronological order) on fluoride concentrations in fingernails and/or toenails

Reference	Exposure	Age range, years	Site	Sample size	Main findings
Buzalaf et al. [15]	chronic	4–6	fingernails/ toenails	121	correlations for F intake and nail concentrations were higher than for intake and urinary excretion (in children exposed to fluoridated water, salt and milk)
Lima-Arsati et al. [54]	subchronic	1–3	fingernails	23	no detectable difference in nail F concentrations when using a 1,500-ppm F dentifrice instead of a placebo toothpaste
Fukushima et al. [13]	chronic	3–7 14–20 30–40 50–60	fingernails/ toenails	300	(1) geographical area and water F concentration exerted the most influence on finger/toenail concentrations; (2) higher F concentrations in nails from older subjects and females
Buzalaf et al. [14]	chronic	5–6	fingernails/ toenails	60	significant differences between F concentrations in nails from children using conventional toothpaste (1,100 ppm, pH 7.0) and low-F toothpaste (550 ppm, pH 4.5)
de Almeida et al. [21]	chronic	1–3	fingernails	33	no correlation between F intake and concentration in nails
Buzalaf et al. [9]	subchronic	20–30	fingernails/ toenails	10	increased F concentrations in toenails 3.7 months after increased F intake (1.8 mg/day, 30 days)
Pessan et al. [55]	subchronic	4–7	fingernails	20	no detectable difference in nail concentrations when using a 1,500-ppm F dentifrice instead of a placebo toothpaste
Levy et al. [12]	chronic	2–6	fingernails/ toenails	30	significant differences in fingernail and toenail F concentrations between children living in areas with different F concentrations in the drinking water
Correa Rodrigues et al. [8]	subchronic	2–3	fingernails/ toenails	10	increased F concentrations in nails when using a 1,500 ppm F dentifrice instead of a placebo toothpaste

Table 1. Continued

Reference	Exposure	Age range, years	Site	Sample size	Main findings
Whitford et al. [7]	(a) chronic	6–7	fingernails	46	(a) significant differences in nail F concentrations among residents in communities with 0.1, 1.6 and 2.3 ppm F in the drinking water
	(b) chronic	adult	fingernails/ toenails	1	(b) F concentrations in fingernails higher than in toenails
	(c) subchronic	adult	fingernails	1	(c) Increased F concentrations in fingernails 3.5 months after increased F intake (3 mg/day, 30 days)
Spate et al. [56]	chronic	adult (women)	toenails	25	higher F concentrations in nails from residents in an area with 1 ppm F in the water when compared to those from a community with 0.1 ppm F
Schmidt and Leuschke [57]	chronic	42–86	fingernails	38	higher F concentrations in nails of individuals exposed to polluted air
Czarnowski and Krechniak [10]	subchronic	21–61	fingernails	110	significant correlation ($r = 0.99$) between F concentrations in nails and urine, which were directly related to exposure to F from the environment
Schamschula et al. [17]	chronic	children	fingernails	139	F concentrations in nails directly proportional to F concentrations in the drinking water of 3 communities
Balazova [16]	chronic	6–14	info. not available	info. not available	higher F concentrations in nails of children living in close proximity of an aluminum smelter in comparison to controls

F = Fluoride. Source: Clarkson et al. [58] (updated).

a higher blood supply. However, as that pattern could not be demonstrated by other investigators, this question remains to be answered. One aspect that must be taken into account, however, is that toenails may be less prone to external contamination by fluoride than fingernails, which might partially explain the lower values found in some studies. Therefore, it has been suggested that toenails, especially from the big toes (which provide enough mass for fluoride analysis) should be used in future studies evaluating fluoride levels in nails [9, 13].

The analytical method used for sample preparation and analysis also seems to be an important variable to be considered [10]. All papers published after the study by Whitford et al. [7] used HMDS (hexamethyldisiloxane)-facilitated diffusion (Taves [18], modified by Whitford [19]), so if the results obtained by those investigators are used as reference values, it can be seen that

Table 2. Multivariate linear regression analysis of factors associated to fingernail fluoride concentration

Variable	Coefficient	SE coeff.	β	SE β	p value	Adjusted R^2
Water fluoride level	1.95	0.33	0.30	0.05	<0.001	0.33
Age	0.03	0.01	0.15	0.05	0.002	
Gender 0 = Female 1 = Male	−2.15	0.41	−0.25	0.05	<0.001	
Geographical area 0 = Southeast (A, C) 1 = Northeast (B, D, E)	3.78	0.55	0.45	0.07	<0.001	
Class 0 = Urban (A, C, D) 1 = Rural (B, E)	−2.43	0.54	−0.29	0.06	<0.001	

The variable growth rate was included in the model, but was removed since it was not statistically significant. A–E = 5 Brazilian communities used in the original study. Source: Fukushima et al. [13].

the results are somewhat comparable, given the differences in the background fluoride exposure (diet and use of dental products). Whitford [2] compared his previous results (Whitford et al. [7]) with those obtained by other investigators using different preparative methods and analytical techniques. Although a clear dose-response relationship could be observed between nail fluoride concentrations and the level of exposure to fluoride in each report, the author observed markedly lower or higher values, showing that care must be taken when evaluating results from investigators using different methodologies. The HMDS method presents some advantages, as samples are not required to be minced or ashed, and cleaning the nails by brushing with deionized water can be done without loss of intrinsic fluoride, reinforcing the concept that fluoride is mainly incorporated into nails from the systemic circulation [7].

In order to provide stronger evidence for the validation of nails as biomarkers of fluoride exposure, a recent study evaluated factors that might influence the fluoride concentrations in fingernails and toenails [13]. The effects of water fluoride concentration, age, gender, nail growth rate and geographical area on the fluoride concentration in the fingernail and toenail clippings were evaluated in 300 volunteers, distributed into 4 age groups. Among the tested factors, geographical area and water F concentration exerted the most influence on finger- and toenail fluoride concentrations (tables 2, 3). Subjects of older age groups from communities located at a warmer region and with naturally fluoridated drinking water showed higher nail fluoride concentrations than the others. Females presented higher nail fluoride concentrations than males. The authors concluded that water fluoride concentration, age, gender and geographical area influenced the fluoride concentrations of fingernails and toenails, and should be taken into account when using this biomarker of exposure to predict risk for dental fluorosis.

Although the use of nails seems to be promising for monitoring fluoride exposure in humans,

Table 3. Multivariate linear regression analysis of factors associated to toenail fluoride concentration

Variable	Coefficient	SE coeff.	β	SE β	p value	Adjusted R^2
Water fluoride level	0.404	0.034	0.484	0.040	<0.001	0.58
Age	0.003	0.001	0.101	0.039	0.009	
Gender 0 = Female 1 = Male	−0.140	0.042	−0.129	0.039	0.001	
Geographical/social- economic area 0 = Southeast (A, C) 1= Northeast (B, D, E)	0.454	0.044	0.419	0.040	<0.001	

The variable growth rate was included in the model but was removed since it was not statistically significant. A–E = 5 Brazilian communities used in the original study. Source: Fukushima et al. [13].

some points need to be addressed. The main issues about the validation of nails as biomarkers of fluoride exposure are related to the sensitivity and specificity of the method. It is possible that nails could only be used to differentiate levels of fluoride exposure when there is a broad variation among them. A clinical study demonstrated significant differences in fluoride concentrations in fingernails among children with Thylstrup-Fejerskov scores of 0 and 5, but not in children with Thylstrup-Fejerskov scores of 1, 2, 3 and 4 [20]. Another study, which evaluated fluoride intake from diet and dentifrice in 1- to 3-year-old children, also demonstrated that small variations in the daily dose of fluoride intake cannot be detected in fingernails, suggesting that fingernails give an indication of fluoride intake over the long term and are unlikely to be sufficiently sensitive to distinguish small day-to-day variations in fluoride intake [21]. Another problem concerning sensitivity and specificity is evidenced by studies which evaluated fluoride concentrations in nails of subjects residing in areas with different levels of exposure to fluoride. Although most of those studies were able to determine significant differences in nails fluoride concentrations among

different populations, it is not uncommon to observe overlaps among the 95% CI in the analyzed populations.

It must be highlighted, however, that no study correlated dental fluorosis severity with fluoride concentrations to which children were exposed during the formation of the permanent dentition, so the sensitivity and specificity of nails to predict dental fluorosis still need to be determined. To date, only one ongoing study conducted with Brazilian children has addressed this issue. Fingernail clippings were collected from children during the period of formation of permanent teeth and analyzed for fluoride content, and this information was later correlated with dental fluorosis prevalence in the permanent dentition (unpublished data). Fluoride concentrations in fingernails of children presenting dental fluorosis were significantly higher than those not presenting fluorosis. More importantly, although some values could be classified as outliers, fluoride concentrations in fingernails were directly related to the severity of dental fluorosis in the majority of the cases. These exciting results, although they should be interpreted with caution, indicate that nails could be used in the near future as indicators of systemic exposure to fluoride that could lead to the development of dental fluorosis.

Another issue involving the use of nails as biomarkers of exposure to fluoride is the possibility of fluoride uptake from exogenous sources. Although Whitford et al. [7] showed no effect of prolonged immersion of nails in deionized or fluoridated water (1 ppm) on the resulting fluoride concentration, a more recent investigation demonstrated that soaking nails in a 1,100-ppm fluoride dentifrice slurry for 3 minutes, or in water with 100 ppm fluoride for 2 h can dramatically increase the fluoride concentration in nails [22]. Vigorous cleaning of nail fragments by brushing with deionized water and sonication were shown to be unable to remove fluoride incorporated from the dentifrice slurry and from the solution, leading to the conclusion that an inappropriate external decontamination can lead to a misclassification of exposure when using nails as a fluoride biomarker. Other disadvantages include: (1) the need of fluoride extraction (in contrast to other biomarkers), which increases both time and cost; (2) the unsuitability of some nails for analysis, such as nails covered with polish (as some products contain fluoride); (3) the possible effects of some variables on the rate of fluoride incorporation into fingernails remain to be determined (as nail diseases).

Considering all the advantages and disadvantages of monitoring chronic and subchronic exposure to fluoride using nails, it is evident that this biomarker is a promising tool to be used in dentistry. Although nail clippings still cannot be used as predictors of dental fluorosis in individual subjects, they have a strong potential for epidemiological surveys, as demonstrated by several reports using different research protocols. More studies are still needed before nails can be fully validated as biomarkers of exposure to fluoride. The appropriate method for decontamination of fluoride from external sources, without removing fluoride incorporated from the blood supply, would be extremely important. Also, the determination of the variables that affect individual variations would be instructive.

Finally, based on the results from the existing literature, future studies should use toenails when monitoring fluoride exposure, and growth rate and length must be taken into account when monitoring subchronic exposure. Also, water fluoride concentration, age, gender and geographical area should be considered when using this biomarker of exposure to predict risk for dental fluorosis.

Hair

As described for nails, the use of hair as an indicator of systemic exposure to fluoride has been the subject of investigations for over 4 decades, both in studies with animals [23, 24] and humans [10, 16, 17, 25–28]. The rationale for the use of hair as a biomarker of fluoride (and other elements) exposure is the same of that for nails: the endogenous trace element composition of hair and nails is believed to reflect the metabolic environment during formation and to be relatively isolated when the finished structure is expelled from the skin [29].

Hair has been reported to be a suitable biomarker to monitor fluoride exposure from different environmental sources. Balazova [16] evaluated fluoride concentrations in the hair of children after the start of operation of an aluminum smelter, and compared these to values obtained in children from a control area. Mean fluoride concentration in the hair of exposed children was 16.0 µg/g, about twice the amount in non-exposed children (7.5 µg/g). Such increases were also observed for fluoride content of teeth, nails and urine of exposed children. According to the author, that study prompted the introduction of measures for the protection of the life and health of the population affected. After 10 years, medical examinations along with analysis of fluoride in teeth, nails and urine were repeated, and the results showed that mean fluoride concentrations in hair had decreased from 16.0 to 4.0 µg/g [30].

Other studies also reported the usefulness of hair in monitoring the levels of exposure to fluoride in hydrofluoric acid workers [26], as well as in phosphate fertilizer workers [10]. In both cases, mean fluoride concentrations in hair of exposed workers were significantly higher than those observed for the control subjects.

Hair has also been shown to be effective in detecting significant differences among children exposed to water containing different fluoride levels (≤0.11 ppm; 0.5–1.1 ppm; 1.6–3.1 ppm) [17]. Mean fluoride concentrations in hair increased consistently and significantly with increasing water fluoride levels, and such a trend was also seen for urine, nails, saliva, enamel and plaque samples. A direct relationship was observed between water fluoride concentration and dental fluorosis, while an inverse relationship was seen between water fluoride concentration and dental caries.

Although the use of hair as biomarker of fluoride exposure started to be investigated at about the same time as nails, there seems to have been a lack of interest in the subject in the last few years. With the exception of one study, the majority of works assessing the use of hair as a biomarker of exposure to fluoride were conducted over 15 years ago. The most recent report addressing the use of hair as indicator of systemic fluoride exposure showed that fluoride content in the drinking water is highly correlated with fluoride content in hair and with dental fluorosis levels in 12-year-old Serbian children [28]. The authors stated that hair might be regarded as a biomaterial of high informative potential in evaluating prolonged exposure to fluorides and to individuate children at risk of fluorosis, regardless of the phase of teeth eruption.

As for nails, despite the promising results mentioned above, the assessment of fluoride exposure using hair presents issues regarding the method of fluoride extraction and external contamination. Ophaug [29] discussed possible factors that could influence the results obtained depending on the method of pretreatment sampling, by

comparing the studies of Schamschula et al. [17] and Czarnowski and Krechniak [10]. In both studies, fluoride was analyzed with a fluoride-selective electrode using the same extraction method, and both studies found a positive correlation between fluoride exposure and the fluoride content of the hair and nails. However, as described by Ophaug [29], an 8-fold increase in systemic fluoride exposure in the study of Schamschula et al. [17] increased the fluoride content of hair and nails by a factor of approximately 2 and 3, respectively; in contrast, a 3-fold increase in systemic fluoride exposure in the study by Czarnowski and Krechniack [10] led to a 200-fold increase in fluoride concentrations in hair, and a 14-fold increase in nail fluoride concentration. In the review by Ophaug [29], issues concerning the length of hair used for analysis, as well as the decontamination procedures used are regarded as the possible causes for the conflicting results obtained.

In addition to issues regarding external contamination of nails, it is worth mentioning that despite the significant differences observed among groups exposed to different sources of fluoride, some individual values overlap between the groups, so no reference value is currently available to indicate safety levels of systemic exposure to fluoride. Also, as hair sampling needs to be done as close to the scalp as possible, it may not be accepted by all subjects, especially those with long hair.

Historical Biomarkers of Exposure to Fluoride

Bone

General characteristics of bone tissue are described by Rugg-Gunn et al. [this vol., pp. 37–51], while the role played by bone in the metabolism of fluoride is described by Buzalaf and Whitford [this vol., pp. 20–36]. Briefly, around 40% of an absorbed amount of fluoride will become incorporated into bone in normal adults and even higher percentages (around 55%) in children [31]. Thus, bone is the main site of fluoride accumulation in the body. Fluoride levels in bone throughout life tend to increase because 99% of the body burden of fluoride is associated with calcified tissues, mainly with the skeleton [32]. This makes bone a natural candidate as a fluoride biomarker.

In order to understand the potential applications of bone as a biomarker of exposure to fluoride, it is important to comprehend in which sites fluoride accumulates within the tissue. A physiologically based pharmacokinetic model considers that bone has two compartments: a small, flow-limited, rapidly exchangeable surface bone compartment, and a bulk, virtually non-exchangeable, inner bone compartment. However, it is known that fluoride associated with the inner bone compartment is not irreversibly bound. In the longer time frame it may be mobilized through the continuous process of bone remodeling in the young, bone resorption and bone remodeling in the adult. These features were elegantly described in a pharmacokinetics study related to chronic exposure to fluoride [33]. It is important to highlight that fluoride is not taken up by fully mineralized bone. It accumulates solely in bone formed during exposure to fluoride. Due to this, incorporation of fluoride in adult bone occurs only during remodeling.

In the 1990s, a mathematical model for fluoride uptake by the skeleton was proposed [34]. This model suggested that: (1) binding of fluoride to bone is nonlinear; smaller percentages of fluoride bind to bone upon higher levels of fluoride intake; (2) bone resorption rate is inversely related to bone fluoride content since it is proportional to the solubility of hydroxyfluorapatite; (3) fluoride clearance from the skeleton by bone remodeling takes over four times longer than fluoride uptake, and, as a consequence, bone fluoride concentrations increase with age [35–39].

Other variables that might affect fluoride incorporation into bones are genetic background [40–43], renal function [44] and remodeling rate [36]. Bone fluoride concentrations also depend on the type of bone analyzed. They are higher in cancellous bone when compared with compact bone due to the higher blood supply of the former [31, 38, 39]. Gender also seems to influence bone fluoride concentrations. Men usually present higher concentrations than women, who appear to have delayed bone fluoride uptake [36, 39]. All these variables make bone fluoride concentrations extremely variable, turning difficult the establishment of a normal range of fluoride levels in bone that would indicate 'desirable' exposure to fluoride. However, it is possible to detect differences in bone fluoride concentrations of individuals living in areas with distinct fluoride concentrations provided that samples matched for age are compared [45–47].

The main issue when attempting to use bone as biomarker of fluoride exposure is the difficulty and invasiveness of sample collection. Most of the studies that evaluated bone fluoride concentrations collected post-mortem samples or samples from subjects undergoing orthopedic surgery, which considerably limits the usefulness of this biomarker.

Teeth
As discussed by Buzalaf and Whitford [this vol., pp. 20–36], mineralized tissues are the main site of fluoride retention in the organism. Thus, they are considered to be biomarkers for total fluoride body burden or historical biomarkers of exposure to fluoride. Despite the fact that most of the fluoride retained in the organism is associated with the skeleton, part of it is deposited in the teeth. Since the collection of bone is difficult and invasive, teeth have emerged as potential historical biomarkers of exposure to fluoride, in particular third molars or premolars that are commonly extracted.

Unlike bone, teeth comprise two distinct types of mineralized tissues: enamel and dentin. The timing and characteristics of fluoride uptake in both tissues are quite different, which has implications for the usefulness of these biomarkers. In enamel, systemic fluoride uptake occurs only during tooth formation. As a consequence, bulk enamel fluoride concentrations mainly reflect the level of systemic exposure to fluoride during tooth formation [32]. However, after tooth eruption, enamel fluoride concentrations are subject to change. Enamel fluoride concentrations are highest at the surface and reduce progressively toward the dentin-enamel junction. Fluoride concentrations at the enamel surface tend to decrease with age in areas subjected to tooth wear [48–50]. On the other hand, they tend to increase in areas that accumulate plaque, since carbonated hydroxiapatite is gradually replaced by fluoridated hydroxiapatite during cariogenic challenges in the presence of topical fluoride [for details, see Buzalaf et al., this vol., pp. 97–114]. The possible changes in enamel fluoride concentrations caused by variables other than fluoride intake (such as wear or sequential de- and remineralization reactions) reduce the ability of this biomarker to estimate fluoride intake or to the predict risk of developing dental fluorosis. Despite one study finding a significant correlation between enamel fluoride content and the degree of dental fluorosis [51], a more recent investigation with a larger sample size evaluating unerupted human third molars from areas with different fluoride levels in the drinking water did not confirm such a correlation [52].

Dentin, in contrast, continues to form and to accumulate fluoride throughout life. Additionally, it only contains fluoride that has been incorporated through systemic ingestion and is protected from topical fluoride exposure by the presence of the covering enamel. Dentin fluoride concentrations, similarly to what has been found for bone and contrarily to enamel, increase with age due to continuous fluoride uptake throughout life [53]. Moreover, they reduce progressively from the pulpal surface to the dentin-enamel junction [48]. The pattern of fluoride uptake in dentin and, consequently, its fluoride concentrations resemble those found for bone [32]. Since dentin is more easily obtained than bone (especially from third molars that are often extracted), it seems to be the best biomarker to estimate chronic fluoride intake and the most appropriate indicator of total fluoride body burden. This was confirmed in a study that found a significant correlation between dentin fluoride concentration and dental fluorosis severity in unerupted third molars of individuals with different levels of exposure to fluoride from the drinking water. However, the coefficient of determination was very low ($r^2 = 0.1$) [52], suggesting that other factors – such as individual genetic variation that affects the susceptibility to dental fluorosis [40] and the fluoride metabolism [41] – may account for dental fluorosis severity. This makes it difficult to establish 'normal' levels of fluoride in dentin, above which excessive exposure to fluoride would be expected to occur.

Conclusion

The knowledge that fluoride controls caries mainly due to its topical effect made it possible to obtain the maximum benefit of this element with a minimum of unwanted effects. Researchers across the globe then turned their attention towards controlling the amount of fluoride intake. Since this is a hard task due to the plethora of available fluoride sources, the use of biomarkers of exposure to this ion gained considerable attention.

Recent and historical biomarkers assess (chronic or subchronic) exposure to fluoride in the medium- and long-term, respectively. Considering the recent biomarkers, from the available knowledge it seems that nails are more suitable for monitoring fluoride exposure, since more information is available and their collection is less prone to questioning by the subjects. As for the historical biomarkers that could indicate total fluoride body burden, bone – despite being studied more – does not seem to be suitable since its collection is difficult and invasive. Dentin is more appropriate in this regard.

It should be emphasized, however, that none of the above-mentioned biomarkers could be used as predictors of dental fluorosis for individual subjects, since 'usual' concentrations of fluoride have not yet been established due to the great variability in levels. Nevertheless, they seem to have the potential to be used in epidemiological surveys.

References

1 McDonnell ST, O'Mullane D, Cronin M, MacCormac C, Kirk J: Relevant factors when considering fingernail clippings as a fluoride biomarker. Community Dent Health 2004;21:19–24.
2 Whitford GM: Monitoring fluoride exposure with fingernail clippings. Schweiz Monatsschr Zahnmed 2005;115: 685–689.
3 Biological markers in environmental health research. Committee on Biological Markers of the National Research Council. Environ Health Perspect 1987; 74:3–9.
4 Grandjean P: Biomarkers in epidemiology. Clin Chem 1995;41:1800–1803.
5 World Health Organization: Fluorides and oral health. Report of a WHO expert committee and oral health status and fluoride use. World Health Organ Tech Rep Ser 1994;846:37.
6 Buzalaf MA, Caroselli EE, Cardoso de Oliveira R, Granjeiro JM, Whitford GM: Nail and bone surface as biomarkers for acute fluoride exposure in rats. J Anal Toxicol 2004;28:249–252.
7 Whitford GM, Sampaio FC, Arneberg P, von der Fehr FR: Fingernail fluoride: a method for monitoring fluoride exposure. Caries Res 1999;33:462–467.
8 Correa Rodrigues MH, de Magalhaes Bastos JR, Rabelo Buzalaf MA: Fingernails and toenails as biomarkers of subchronic exposure to fluoride from dentifrice in 2- to 3-year-old children. Caries Res 2004;38:109–114.
9 Buzalaf MA, Pessan JP, Alves KM: Influence of growth rate and length on fluoride detection in human nails. Caries Res 2006;40:231–238.
10 Czarnowski W, Krechniak J: Fluoride in the urine, hair, and nails of phosphate fertiliser workers. Br J Ind Med 1990;47: 349–351.
11 Feskanich D, Owusu W, Hunter DJ, Willett W, Ascherio A, Spiegelman D, Morris S, Spate VL, Colditz G: Use of toenail fluoride levels as an indicator for the risk of hip and forearm fractures in women. Epidemiology 1998;9:412–416.

12 Levy FM, Bastos JR, Buzalaf MA: Nails as biomarkers of fluoride in children of fluoridated communities. J Dent Child (Chic) 2004;71:121–125.
13 Fukushima R, Rigolizzo DS, Maia LP, Sampaio FC, Lauris JR, Buzalaf MA: Environmental and individual factors associated with nail fluoride concentration. Caries Res 2009;43:147–154.
14 Buzalaf MA, Vilhena FV, Iano FG, Grizzo L, Pessan JP, Sampaio FC, Oliveira RC: The effect of different fluoride concentrations and pH of dentifrices on plaque and nail fluoride levels in young children. Caries Res 2009;43:142–146.
15 Buzalaf MAR, Rodrigues MHC, Pessan JP, Leite AL, Arana A, Villena RS, Forte FD, Sampaio FC: Biomarkers of fluoride in children exposed to different sources of systemic fluoride. J Dent Res 2011;90: 215–219.
16 Balazova G: Long-term effect of fluoride emission upon children. Fluoride 1971; 4:85–88.
17 Schamschula RG, Sugar E, Un PS, Toth K, Barmes DE, Adkins BL: Physiological indicators of fluoride exposure and utilization: an epidemiological study. Community Dent Oral Epidemiol 1985;13: 104–107.
18 Taves DR: Separation of fluoride by rapid diffusion using hexamethyldisiloxane. Talanta 1968;15:969–974.
19 Whitford GM: The Metabolism and Toxicity of Fluoride, ed 2. Basel, Karger, 1996.
20 Sampaio FC, Whitford GM, Arneberg P, von der Fehr FR: Validation of fingernail fluoride as a biomarker for dental fluorosis (abstract 72). Caries Res 2003;37: 291–292.
21 de Almeida BS, da Silva Cardoso VE, Buzalaf MA: Fluoride ingestion from toothpaste and diet in 1- to 3-year-old Brazilian children. Community Dent Oral Epidemiol 2007;35:53–63.
22 Fukushima R, Sampaio FC, Buzalaf MAR: Risk assessment of external contamination of nails with fluoride (abstract 146). Caries Res 2008;42:235.

23 Krechniak J: Fluorides in drinking water and hair. Fluoride 1975;8:38–40.
24 Zakrzewska H, Brzezińska M, Orowicz W, Samujło D, Wójcik A, Kolanus A: Content of fluoride in hair and hoofs of wild boars and deer from Western Pomerania as a bioindicator of environmental pollution. Ann Acad Med Stetin 2004;50(suppl 1):100–103.
25 Schamschula RG, Duppenthaler JL, Sugar E, Un PS, Toth K, Barmes DE: Fluoride intake and utilization by Hungarian children: associations and interrelationships. Acta Physiol Hung 1988;72:253–261.
26 Kono K, Yoshida Y, Watanabe M, Orita Y, Dote T, Bessho Y: Urine, serum and hair monitoring of hydrofluoric acid workers. Int Arch Occup Environ Health 1993;65:S95–S98.
27 Kono K, Yoshida Y, Watanabe M, Watanabe H, Inoue S, Murao M, Doi K: Elemental analysis of hair among hydrofluoric acid exposed workers. Int Arch Occup Environ Health 1990;62:85–88.
28 Mandinic Z, Curcic M, Antonijevic B, Carevic M, Mandic J, Djukic-Cosic D, Lekic CP: Fluoride in drinking water and dental fluorosis. Sci Total Environ 2010; 408:3507–3512.
29 Ophaug R: Determination of fluorine in biological materials: reaction paper. Adv Dent Res 1994;8:87–91.
30 Balazova G, Lipkova V: Evaluation of some health parameters in children in the vicinity of an aluminium factory. Fluoride 1974;7:88–93.
31 Villa A, Anabalon M, Zohouri V, Maguire A, Franco AM, Rugg-Gunn A: Relationships between fluoride intake, urinary fluoride excretion and fluoride retention in children and adults: an analysis of available data. Caries Res 2010;44:60–68.
32 Whitford GM: Intake and metabolism of fluoride. Adv Dent Res 1994;8:5–14.
33 Rao HV, Beliles RP, Whitford GM, Turner CH: A physiologically based pharmacokinetic model for fluoride uptake by bone. Regul Toxicol Pharmacol 1995;22:30–42.

34 Turner CH, Boivin G, Meunier PJ: A mathematical model for fluoride uptake by the skeleton. Calcif Tissue Int 1993; 52:130–138.

35 Richards A, Mosekilde L, Sogaard CH: Normal age-related changes in fluoride content of vertebral trabecular bone – relation to bone quality. Bone 1994;15: 21–26.

36 Ishiguro K, Nakagaki H, Tsuboi S, et al.: Distribution of fluoride in cortical bone of human rib. Calcif Tissue Int 1993;52: 278–282.

37 Parkins FM, Tinanoff N, Moutinho M, Anstey MB, Waziri MH: Relationships of human plasma fluoride and bone fluoride to age. Calcif Tissue Res 1974;16: 335–338.

38 Eble DM, Deaton TG, Wilson FC Jr, Bawden JW: Fluoride concentrations in human and rat bone. J Public Health Dent 1992;52:288–291.

39 Suzuki Y: The normal levels of fluorine in the bone tissue of Japanese subjects. Tohoku J Exp Med 1979;129:327–336.

40 Everett ET, McHenry MA, Reynolds N, Eggertsson H, Sullivan J, Kantmann C, Martinez-Mier EA, Warrick JM, Stookey GK: Dental fluorosis: variability among different inbred mouse strains. J Dent Res 2002;81:794–798.

41 Carvalho JG, Leite AL, Yan D, Everett ET, Whitford GM, Buzalaf MA: Influence of genetic background on fluoride metabolism in mice. J Dent Res 2009;88: 1054–1058.

42 Mousny M, Banse X, Wise L, Everett ET, Hancock R, Vieth R, Devogelaer JP, Grynpas MD: The genetic influence on bone susceptibility to fluoride. Bone 2006;39:1283–1289.

43 Dequeker J, Declerck K: Fluor in the treatment of osteoporosis: an overview of thirty years clinical research. Schweiz Med Wochenschr 1993;123:2228–2234.

44 Ekstrand J, Spak CJ: Fluoride pharmacokinetics: its implications in the fluoride treatment of osteoporosis. J Bone Miner Res 1990;5(suppl 1):S53–S61.

45 Arnala I, Alhava EM, Kauranen P: Effects of fluoride on bone in Finland: histomorphometry of cadaver bone from low and high fluoride areas. Acta Orthop Scand 1985;56:161–166.

46 Lan CF, Lin IF, Wang SJ: Fluoride in drinking water and the bone mineral density of women in Taiwan. Int J Epidemiol 1995;24:1182–1187.

47 Chachra D, Limeback H, Willett TL, Grynpas MD: The long-term effects of water fluoridation on the human skeleton. J Dent Res 2010;89:1219–1223.

48 Weatherell JA: Uptake and distribution of fluoride in bones and teeth and the development of fluorosis; in Barltrop W, Burland BL (eds): MIneral Metabolism in Paediatrics. Oxford, Blackwell, 1969, pp 53–70.

49 Weatherell JA, Hallsworth AS, Robinson C: Fluoride in the labial surface of permanent anterior teeth. Caries Res 1972; 6:67–68.

50 Weatherell JA, Robinson C, Hallsworth AS: Changes in the fluoride concentration of the labial enamel surface with age. Caries Res 1972;6:312–324.

51 Richards A, Likimani S, Baelum V, Fejerskov O: Fluoride concentrations in unerupted fluorotic human enamel. Caries Res 1992;26:328–332.

52 Vieira AP, Hancock R, Limeback H, Maia R, Grynpas MD: Is fluoride concentration in dentin and enamel a good indicator of dental fluorosis? J Dent Res 2004; 83:76–80.

53 Nakagaki H, Koyama Y, Sakakibara Y, Weatherell JA, Robinson C: Distribution of fluoride across human dental enamel, dentine and cementum. Arch Oral Biol 1987;32:651–654.

54 Lima-Arsati YB, Martins CC, Rocha LA, Cury JA: Fingernail may not be a reliable biomarker of fluoride body burden from dentifrice. Braz Dent J 2010;21:91–97.

55 Pessan JP, Pin ML, Martinhon CC, de Silva SM, Granjeiro JM, Buzalaf MA: Analysis of fingernails and urine as biomarkers of fluoride exposure from dentifrice and varnish in 4- to 7-year-old children. Caries Res 2005;39:363–370.

56 Spate VL, Morris JS, Baskett CK, et al.: Determination of fluoride in human nails via cyclic instrumental neutron activation analysis. J Radioanalyt Nucl Chem 1994;179:27–33.

57 Schmidt CW, Leuschke W: Fluoride content in fingernails of individuals with and without chronic fluoride exposure. Fluoride 1990;23:79–82.

58 Clarkson J, Watt RG, Rugg-Gunn AJ, et al: Proceedings: 9th World Congress on Preventive Dentistry (WCPD): 'Community Participation and Global Alliances for Lifelong Oral Health for All,' Phuket, Thailand, September 7–10, 2009. Adv Dent Res 2010;22:2–30.

Prof. Juliano Pelim Pessan
Department of Pediatric Dentistry and Public Health
Araçatuba Dental School, São Paulo State University
Rua José Bonifácio, 1193
16015–050 Araçatuba – SP (Brazil)
Tel. +55 18 3636 3274, E-Mail jpessan@foa.unesp.br

Buzalaf MAR (ed): Fluoride and the Oral Environment.
Monogr Oral Sci. Basel, Karger, 2011, vol 22, pp 66–80

Acute Toxicity of Ingested Fluoride

Gary Milton Whitford

Department of Oral Biology, Medical College of Georgia, Augusta, Ga., USA

Abstract

This chapter discusses the characteristics and treatment of acute fluoride toxicity as well as the most common sources of overexposure, the doses that cause acute toxicity, and factors that can influence the clinical outcome. Cases of serious systemic toxicity and fatalities due to acute exposures are now rare, but overexposures causing toxic signs and symptoms are not. The clinical course of systemic toxicity from ingested fluoride begins with gastric signs and symptoms, and can develop with alarming rapidity. Treatment involves minimizing absorption by administering a solution containing calcium, monitoring and managing plasma calcium and potassium concentrations, acid-base status, and supporting vital functions. Approximately 30,000 calls to US poison control centers concerning acute exposures in children are made each year, most of which involve temporary gastrointestinal effects, but others require medical treatment. The most common sources of acute overexposures today are dental products – particularly dentifrices because of their relatively high fluoride concentrations, pleasant flavors, and their presence in non-secure locations in most homes. For example, ingestion of only 1.8 ounces of a standard fluoridated dentifrice (900–1,100 mg/kg) by a 10-kg child delivers enough fluoride to reach the 'probably toxic dose' (5 mg/kg body weight). Factors that may influence the clinical course of an overexposure include the chemical compound (e.g. NaF, MFP, etc.), the age and acid-base status of the individual, and the elapsed time between exposure and the initiation of treatment. While fluoride has well-established beneficial dental effects and cases of serious toxicity are now rare, the potential for toxicity requires that fluoride-containing materials be handled and stored with the respect they deserve.

As is true of virtually all substances to which humans are exposed, including water, oxygen and table salt, exposure to high amounts of fluoride can cause adverse effects. It is a toxicological axiom that such effects are due to the level of exposure to the substance, not to the substance itself. Compared to the first half of the 20th century, cases of serious fluoride toxicity are uncommon today. At that time, sodium fluoride was used as a pesticide and rat poison and commonly found in homes, hospitals and elsewhere. Because sodium fluoride powder resembles flour, powdered sugar, baking powder, sodium bicarbonate and similar products, there were many accidental poisonings. Sodium fluoride was also used in a large number of suicides [1].

Table 1. Details of 3 deaths caused by ingestion of fluoride-containing dental products

Age	Body weight, kg	Sex	Dose, mg/kg	Comment	Reference
27 months	not reported	M	3.1–4.5[1]	ingested <100 0.5-mg fluoride tablets; death occurred 5 days later	Dukes [8]
3 years	12.5	M	16	ingested <200 1.0-mg fluoride tablets; vomited; death occurred 7 h later	Eichler et al. [7]
3 years	not reported	M	24–35[1]	swallowed 4% SnF_2 rinse solution; vomited; death occurred 3 h later	Church [6]

[1] Calculated using the 3rd and 97th percentiles for body weight of 3-year-old boys.

One of the most remarkable accidental poisonings occurred at the Oregon State Hospital [2]. About 10 gallons of scrambled eggs were mistakenly prepared with 17 pounds of sodium fluoride instead of powdered milk. There were 263 cases of acute poisoning, of which 47 were fatal. It was not possible to estimate the amounts of fluoride that were ingested by those affected, but the well-known signs and symptoms developed rapidly. Extremely severe nausea, bloody vomiting and diarrhea occurred almost immediately. General collapse accompanied by pallor, weakness, shallow breathing, weak heart sounds, wet cold skin, cyanosis and equally dilated pupils soon followed. When these signs were pronounced, death almost always occurred within 2–4 h. When death was delayed for up to 20 hours, muscle paralysis, carpopedal spasm and spasm of the extremities occurred. More recent reports of serious acute toxicity have indicated that muscular and cardiovascular problems are related to electrolyte imbalances, particularly severe hypocalcemia and hyperkalemia [3, 4]. A progressive, mixed respiratory and metabolic acidosis develops as kidney function and respiration fail.

Doses Causing Serious Toxicity: Rationale for the Probably Toxic Dose

Based on the sketchy information that could be gathered after the mass poisoning at the Oregon State Hospital, Lidbeck et al. [2] thought the acute lethal dose of fluoride was over 100 mg/kg. Hodge and Smith [1] estimated that the 'certainly lethal dose' was between 32 and 64 mg/kg or 5–10 g of sodium fluoride for a 70-kg person. Dreisbach [5] estimated the lethal dose at 6–9 mg/kg. These different estimates probably were due in large part to uncertainty about the amounts of fluoride that were actually ingested by the victims.

Church [6], Eichler et al. [7] and Dukes [8] did not attempt to estimate the acute lethal dose, but did present dosages in their case reports (table 1). The case described by Dukes [8] was unusual because of the small dose and the length of time from ingestion to death. A 27-month-old child ingested an unknown number (but fewer than 100) of 0.5-mg fluoride tablets and experienced respiratory failure. With treatment, the boy's condition improved, but he died 5 days after ingesting the tablets. The amount of fluoride ingested was approximately 50 mg, and the dose was estimated at between 3.1 and 4.5 mg/kg. A 3-year-old boy died in a hospital after ingestion of about 200 1.0-mg fluoride tablets [7]. This child vomited immediately, seemed to recover completely, but then collapsed and died 7 h after the swallowing the tablets. The ingested dose was approximately 16 mg/kg. The case described by Church [6] was a 3-year-old boy who swallowed 4% stannous fluoride rinse from a small cup in a dental clinic. The child vomited immediately, had a convulsive seizure and died 3 h later. The ingested dose was estimated at between 24 and 35 mg/kg, but the absorbed dose was lower because of vomiting.

The report by Eichler et al. [7] also described 108 non-fatal cases of fluoride toxicity in children, most of whom had ingested fluoride tablets. As the ingested dose increased from less than 0.5 mg/kg to 'more than 5.0 mg/kg', the percentage of patients with symptoms increased from 15 to 71. The symptoms included nausea, vomiting and fatigue. One child died as described above. *Based on this report and those of Dukes [8] and Bayless and Tinanoff [9], it is concluded that the probably toxic dose (PTD) for fluoride is 5.0 mg/kg. The PTD is defined as the minimum dose that could cause serious or life-threatening systemic signs and symptoms and that should trigger immediate therapeutic intervention and hospitalization.* This does *not* mean that doses lower than 5.0 mg/kg should be regarded as innocuous.

Treatment

The treatment for serious or potentially life-threatening cases of acute fluoride toxicity must attempt to minimize absorption from the GI tract, increase urinary excretion and maintain the vital signs within levels compatible with life [1, 3, 4, 9]. If vomiting has not occurred, it should be induced unless the patient is unconscious (to avoid aspiration into the lungs). Because of the strong affinity of calcium for fluoride, absorption can be slowed and reduced by the oral administration of 1% calcium chloride or calcium gluconate or, if these solutions are not available, as much milk as can be tolerated. These actions should be taken as soon as possible because fluoride is rapidly absorbed from the stomach and intestines. At the same time, the hospital should be informed that a case of fluoride toxicity is in progress so that appropriate therapeutic interventions are in place when the patient arrives. Expeditious treatment is essential because severe cases often progress rapidly toward death.

In cases of life-threatening toxicity, which must be judged by the clinical signs and symptoms because the exact amount of fluoride ingested, i.e. the dose, is almost never known, an airway and intravenous line should be established immediately upon arrival at the hospital. Blood samples should be obtained upon arrival and then hourly for the measurement of serum fluoride, blood pH and gases, and serum chemistry – including calcium and potassium in particular. Intravenous administration of calcium gluconate to prevent hypocalcemia, glucose to reverse hyperkalemia, and sodium lactate or sodium bicarbonate to minimize acidosis and increase urinary flow and pH in order to increase the urinary excretion rate of fluoride should be given as required. Oxygen therapy, artificial respiration, electrocardiac conversion and hemodialysis may be required. These various measurements and treatments should continue until the vital signs have stabilized and the serum chemistry values have normalized for at least 24–48 h.

Sources of Fluoride

As discussed above, the acute dose of fluoride that may cause serious systemic toxicity is 5 mg/kg (11 mg/kg of sodium fluoride). This is called the 'probably toxic dose' (PTD). It is obvious that optimally fluoridated water (ca. 1.0 mg/l) cannot cause acute toxicity since 5 liters of water would have to be ingested for every kg of body weight.

Table 3. Fluoride contents of dental products and their relationships to the PTD

Product	Concentration of salt fluoride			Amount of product and fluoride usually used		Amount of product containing the PTD for child weighing	
	%	%	ppm	product	fluoride	10 kg	20 kg
Mouthwash							
NaF	0.05	0.023	230	10 ml	2.3 mg	215 ml	430 ml
NaF	0.20	0.091	910	10 ml	9.1 mg	55 ml	110 ml
SnF$_2$	0.40	0.097	970	10 ml	9.7 mg	50 ml	100 ml
Dentifrice							
NaF	0.22	0.10	1,000	1 g	1.0 mg	50 g	100 g
MFP	0.76	0.10	1,000	1 g	1.0 mg	50 g	100 g
Topical gel							
NaF (APF, tray)	2.72	1.23	12,300	5 ml	61.5 mg	4 ml	8 ml
SnF$_2$ (brush)	0.40	0.097	970	1 ml	0.97 mg	50 ml	100 ml
NaF tablet							
0.25 mg	–	–	–	1/day	0.25 mg	200 tabs	400 tabs
0.50 mg	–	–	–	1/day	0.50 mg	100 tabs	200 tabs
1.00 mg	–	–	–	1/day	1.00 mg	50 tabs	100 tabs

PTD = 5 mg/kg i.e. the amount of ingested fluoride that could cause serious or life-threatening systemic effects and that should trigger immediate therapeutic intervention and hospitalization; MFP = sodium monofluorophosphate; APF = acidulated phosphate fluoride. The average body weights of 1-year-old and 6-year-old children are approximately 10 and 20 kg, respectively.

in the USA are required by the Food and Drug Administration to specify that children under the age of 6 years should not use a fluoride mouthwash, they do have access to them in many homes. Dentifrices are available in tubes containing up to 8.2 ounces (232 g) so a 1,000-ppm product contains 232 mg of fluoride. Ingestion of only 1.76 ounces (50 g) by a 10-kg child provides enough fluoride to reach the PTD. As for fluoride tablets, the American Dental Association guidelines state that up to 480 0.25-mg tablets, 240 0.50-mg tablets, and 120 1.0-mg tablets may be prescribed per household [13]. These numbers of tablets and amounts of fluoride contained in them exceed the PTD for both 10-kg and 20-kg children.

All of these products should be kept out of the reach of small children and secured with child resistant caps.

Topical acidulated phosphate fluoride (APF) gels and foams contain fluoride at a concentration of 12.3 mg/ml (12,300 ppm). APF treatments are rarely given to 1-year-old children, but they may be given to 2-year-old children (average body weight 12.4 kg). If maxillary and mandibular stock trays are loaded with 5 ml of gel in each tray, then 123 mg of fluoride would be placed in the mouth, which exceeds the PTD for a 2-year-old by a factor of 2, so swallowing one half of the applied gel would reach the PTD. The currently recommended procedure for APF gel treatments minimizes the amount of gel that is likely to be swallowed and it should be followed. The recommendations are: (1) use the minimum amount of gel required to cover the teeth; (2) use no more than 2 ml of gel in each stock tray; (3) if custom-made trays are used, then use only 5–10 drops of gel in each tray; (4) seat the child in an upright position with the head inclined slightly forward to discourage swallowing; (5) use a saliva ejector throughout the procedure; and (6) allow the child to expectorate for 30 s after the procedure. When this procedure is used, the risk of even temporary stomach irritation due to swallowing is minimal. It is also worth noting that, while the APF foams have the same fluoride concentration as the gels, much less fluoride is placed in the patient's mouth because much of the volume is occupied by air [14].

Whitford et al. [15] reported that, when in an acidic solution, the threshold fluoride concentration that produces histological and functional damage to the canine stomach mucosa is between 19 and 95 mg/l or 1.0 and 5.0 mmol/l. It should be noted that the pH of APF products is approximately 3.5. The pK of hydrofluoric acid (HF) is 3.4, so nearly 50% of the fluoride in the gel or foam exists as undissociated HF (ca. 6,000 mg/l), a molecule that is very irritating to the stomach mucosa and at a concentration far above that known to damage the stomach mucosa. Unless care is taken to reduce swallowing even small amounts, nausea and vomiting may occur.

Products intended for self-application at home may also cause damage to the stomach. Spak et al. [16] examined the histological effects of a 0.42% fluoride gel (4,200 mg/l) with a pH of 6.5. Ten subjects added 1.5 g of the gel to each custom-made maxillary and mandibular tray (a total of 12.6 mg of fluoride) for a 5-min topical application. The average amount of fluoride not recovered from the mouth was 5.1 mg or 40% of the amount applied which was due to using more than the recommended volume for custom-made trays. None of the subjects experienced nausea,

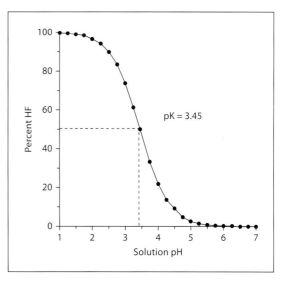

Fig. 1. Relationship between the pH of a solution and the percentage of fluoride that exists as HF.

but endoscopic examination 2 h after the gel treatment revealed mucosal petechiae or erosions. Biopsies of the mucosa showed histological changes, including dilation of the gastric pits, localized losses of surface epithelium and bleeding, in 9 of the 10 subjects.

Effects on the Stomach

Following the ingestion of a large amount of fluoride, or a relatively small amount in a small volume (i.e. a high concentration), the first organ to be affected is the stomach. The contents of the stomach have a distinctly acidic pH. Between meals the pH is usually between 2 and 4, while during and for 1–2 h after meals it is between 1 and 2. Figure 1 shows the relationship between the pH of a solution and the percentage of the total amount of fluoride in the solution that is combined with hydrogen ions to form HF, a weak acid whose pK_a is 3.45. The Henderson-Hasselbalch equation is used to calculate the relative concentrations of ionic fluoride and HF at different pH values. The equation is:

$$pH = pK + \log([F^-]/[HF])$$

When the pH of the stomach contents is 4.0, 22% of the fluoride is in the form of HF. When the pH is lower than 2.0, more than 95% is in the form of HF. HF is a highly diffusible and permeating molecule that diffuses down its concentration gradient to cross cell membranes and epithelia including the stomach mucosa, a tissue that is relatively impermeable to most other ingested substances [17]. Upon entering the gastric mucosa where the pH is close to neutrality, HF dissociates immediately to release fluoride and hydrogen ions. In sufficiently high concentrations, these ions can disrupt the structure and function of the stomach [18–20].

It is important to understand that the effects of fluoride on the stomach are dependent on the *concentration* of fluoride (actually the HF concentration as discussed below) in contact with the mucosa, not on the ingested dose (i.e. mg/kg). For example, if 0.5 liters of water containing 5 mg of fluoride (10 ppm or 0.5 mmol/l) were ingested, it is unlikely that even the mildest of symptoms would be felt and there would be only minimal or no adverse effects on the stomach. If, however, 5 ml of water containing 5 mg of fluoride (1,000 ppm or 52.6 mmol/l) were ingested, nausea and perhaps vomiting and dizziness would be experienced by many people. In each case the same amount of fluoride would have been ingested, but the effects would be quite different. In fact, following the systemic absorption of fluoride, the toxic effects on internal organs are also dependent on the tissue concentration of fluoride but, because the concentrations in these organs are almost never known, the dose for systemic effects is usually expressed in terms of body weight (mg/kg).

The effects of pH on the gastric effects of fluoride were tested using in situ experiments with dogs [21] (fig. 2). Through a midline abdominal incision, a portion of the stomach from the greater curvature with its gastrosplenic blood supply intact was mounted in a two-compartment Lucite chamber with the mucosal surface facing upwards as described by Moody and Durbin [22]. The septum of the chamber divided the tissue into two halves of equal surface area (14.2 cm^2) so that control and test solutions could be placed side-by-side on the mucosa. This permitted a comparison of ion fluxes across the mucosa as well as direct observation of any gross changes that might occur.

The mucosa on one side of the chamber was exposed to a saline solution with or without 10 mmol/l sodium fluoride (190 ppm) at a pH of 6.2 for 22 15-min collection periods. At this pH, only 0.2% of the fluoride is in the form of HF. The mucosa on the other side was exposed to a saline solution acidified with 0.1 N HCl (pH 1.6) also with or without 10 mmol/l sodium fluoride. At this pH, 98.6% of the fluoride is in the form of HF. The solutions without fluoride served as the negative control solutions. The fluxes of water (determined by changes in the concentration of ^{14}C-inulin) and sodium, potassium and hydrogen ions were not affected by 10 mmol/l fluoride when the solution pH was 6.2 and the gross appearance of mucosa remained normal throughout the 5.5-hour study.

In contrast, the water and ion fluxes increased immediately upon exposure to the pH-1.6 solution containing 10 mmol/l fluoride. The water, sodium and potassium fluxes were positive, i.e. they were directed from the mucosa into the test solution. The hydrogen ion fluxes, however, were negative, i.e. directed into the mucosa, which indicated that HF was diffusing down its concentration gradient from the solution in the chamber and thus carrying hydrogen and fluoride ions into the tissue. When the mucosa was repeatedly exposed to the control solution (without fluoride) after the exposures to 10 mmol/l fluoride, the fluxes were reduced slightly but they did not

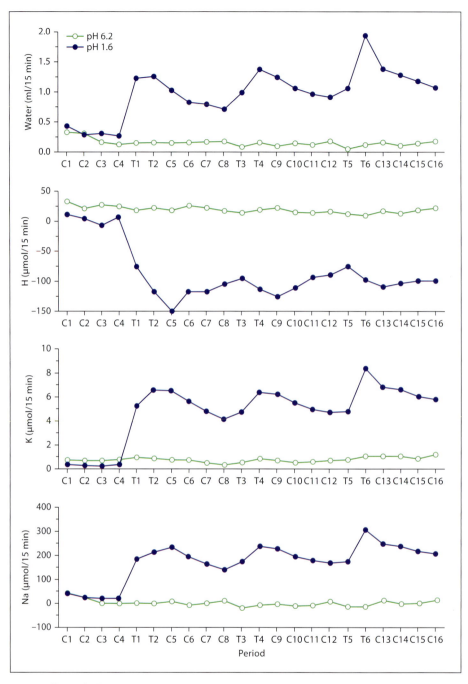

Fig. 2. Effects of solution pH (6.2 or 1.6) on water and ion fluxes from the gastric mucosa of the dog in response to exposure to sodium fluoride. The control solutions (labeled C1–C16) contained no fluoride. The test solutions (labeled T1–T6) contained 10 mmol/l sodium fluoride. Fresh solutions were placed on the mucosa every 15 min.

return to the baseline values. Further, there was an obvious increase in the secretion of mucus followed by swelling (edema) and localized areas of hemorrhage within the first few minutes after placing the pH-1.6 solution containing fluoride on the mucosa. These findings made it clear that changes in the structure and function of the gastric mucosa are caused by exposure to high concentrations of HF, and that equally high concentrations of ionic fluoride are without such effects.

Using the same model, experiments were done to determine the threshold HF concentration for gastric toxicity [15]. The solution used on the control side contained 50 mmol/l sodium chloride in 0.1 N HCl (pH 1.6). The same solution was used on the test side but also contained fluoride in the form of HF at 1.0, 5.0 or 10.0 mmol/l. The water and ion fluxes throughout the 4-hour study (16 15-min collection periods) and the gross and histological appearances on the control side were normal. On the test side, exposure to the solution containing 1.0 mmol/l fluoride as HF produced small but statistically non-significant increases in the fluxes and only minor changes in the appearance of the mucosa. However, all fluxes increased significantly upon exposure to the solution containing 5.0 mmol/l HF. Mucus secretion increased as did the redness and swelling of the mucosa. Subsequent exposure to the 10.0 mmol/l HF solution caused these effects to increase. Upon microscopic examination, the thickness of the surface mucus layer and the epithelium were greatly reduced. In some sections, evidence of surface cell exfoliation was seen indicating cell degeneration and necrosis. It was concluded that the threshold concentration for adverse effects of fluoride in a strongly acidic solution, i.e. HF, is more than 1.0 mmol/l (19 ppm) but less than 5.0 mmol/l (95 ppm). This explains why swallowed APF gel is damaging to the gastric mucosa and should be avoided. The total fluoride concentration (i.e. ionic fluoride plus HF) is 1.23% or 647 mmol/l (12,300 ppm) and, at pH 3.5, the HF concentration is 305 mmol/l (6,104 ppm).

Factors That Influence Toxic Effects

Chemical Compound

The compounds of fluoride vary greatly with respect to their solubilities. Very insoluble compounds such as calcium fluoride, cryolite (Na_3AlF_6), hydroxyfluorapatite and fluorapatite are poorly absorbed from the GI tract. Because of this their LD_{50} values, as determined in studies with laboratory animals, are much higher than those of highly soluble compounds such as sodium fluoride, fluorosilicic acid (H_2SiF_6) and sodium fluorosilicate (Na_2SiF_6), the three compounds that are commonly used to fluoridate drinking water at low fluoride concentrations.

The highest fluoride concentrations to which most people are regularly exposed are found in certain dental products, particularly dentifrices which typically contain 1,000–1,500 mg /kg fluoride. The compounds most often added to dentifrices are sodium fluoride and disodium monofluorophosphate or MFP (Na_2PO_3F). The fluoride in MFP is covalently bonded to the phosphorus. Its release from MFP is slow in water and dentifrices, but rapid in the presence of phosphatases found in the intestine, plasma and internal organs [23]. This was demonstrated in an experiment with 2 groups of rats that were given fluoride intravenously (2.0 mg/kg) as sodium fluoride or MFP [24]. Three blood samples were collected at 10, 30 and 60 min after administration of the doses. The plasma fluoride concentrations in the 2 groups were virtually identical, which indicated the complete hydrolysis of fluoride from MFP prior to the 10-min blood collections.

Based on the time courses of plasma concentrations, however, there is evidence that the absorption of orally administered fluoride when given as MFP is somewhat slower than that from sodium fluoride [23]. The delayed absorption and lower peak plasma fluoride concentrations appear to be due to the limited amount of phosphatase activity in the stomach compared to the intestine. There are, however, no significant differences in the percentages of the doses that are ultimately absorbed systemically. The limited phosphatase activity in the stomach was also indicated in a study with humans by Müller et al. [25]. The subjects ingested sodium fluoride or monofluorophosphate tablets for 1 week. The gastric mucosa was then examined with a gastroscope. No significant damage was found in the MFP group, but acute hemorrhages and free blood were found in the NaF group.

The relative absence of gastric phosphatase activity also explained the lack of functional and structural effects of MFP on the canine mucosa [21]. Using the same split-chamber method described above, the mucosa on one side of the chamber was exposed to 10 mmol/l F as NaF and to 10 mmol/l MFP on the other side for two 15-min periods. The immediate effects on the NaF side included large increases in the fluxes of water, sodium and potassium, increased mucus secretion and increased mucosal swelling and redness. None of these effects occurred on the MFP side except for a slight and transient increase in the potassium flux.

Theoretically, the lower peak plasma fluoride concentrations could reduce the acute toxicity of MFP compared to sodium fluoride. In their study with rats, Shourie et al. [26] reported that the 24-h LD_{50} doses for these two compounds were 75 and 36 mg/kg, respectively. In their study with mice, Lim et al. [27] reported LD_{50} values of 94 and 44 mg/kg, respectively. These findings were used to support the increase of the total fluoride amount as MFP above the limit established by the American Dental Association (260 mg total fluoride per tube of dentifrice). More recent studies, however, could not confirm such differences. In their study with rats, Gruninger et al. [28] reported LD_{50} values of 102 and 98 mg/kg for MFP and sodium fluoride – with mice the values were 54 and 58 mg/kg, respectively. In their study with rats, Whitford et al. [29] reported LD_{50} values of 84.3 and 85.5 mg/kg for MFP and sodium fluoride. Based on these results, the authors stated

that '. . .professional organizations and regulatory agencies should not endorse the policy of adding greater amounts of fluoride, as MFP, to dental products based on the concept that fluoride in the form of MFP is less hazardous than that in the form of NaF'.

Age
Maynard et al. [30] and Mornstad [31] reported that, compared to adult laboratory animals, young laboratory animals are more resistant to the acute toxic effects of fluoride. It is not known whether this is true for humans but there is reason to think that it is. As mentioned earlier, the systemic effects of acute exposures to high doses of fluoride are directly related to the concentrations in plasma and the target organs. The rate of removal of fluoride from plasma and the target organs depends almost entirely on the rates of uptake by calcified tissues, which contain 99% of the fluoride in the body, and excretion in the urine. Therefore, any factor that increases these rates should reduce the severity of the acute toxic effects.

Miller and Phillips [32] fed 3 groups of rats a diet with the same fluoride concentration for 4.5 months. The rats in 1 group began consuming the diet when they were weaned (21 days of age) while 2 other groups started at 9 or 20 weeks of age. At the end of the 4.5-month feeding periods, the bone fluoride concentrations were inversely related to the age at which the rats entered the study – the younger rats had higher concentrations. Similar results were reported by Zipkin and McClure [33] and Suttie and Philips [34] who used rats and by Weidmann and Weatherell [35] who used rabbits.

Whitford [21] used dogs of different ages (4 weeks, 6 months and several years) and infused isotonic solutions containing sodium fluoride intravenously for 20 min and then the infusion pump was turned off. Blood samples were collected 12 times over 6 h. Each dog received the same dose in terms of body weight (5.0 mg/kg). The

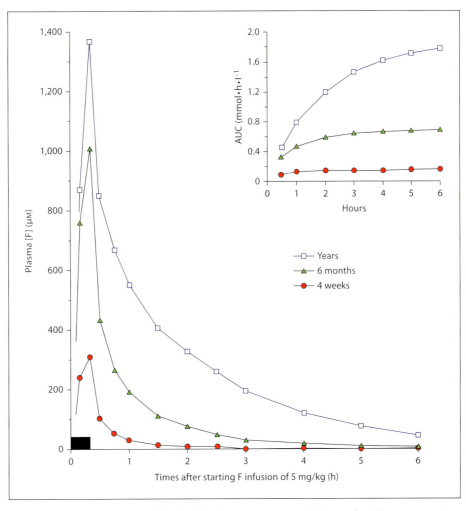

Fig. 3. Effect of age, or stage of skeletal development, on arterial plasma fluoride concentrations in dogs. The cumulative areas under the plasma-time curves (AUC) are shown [21].

peak plasma fluoride concentrations were 305, 1,004 and 1,367 µmol/l, respectively, and the areas under the time-plasma concentration curves were markedly higher in the older animals as well (fig. 3). Similar results were found in a study with rats that were 23 days or 6.5 months of age [21].

These age-related differences in plasma fluoride concentrations were due almost entirely to a greater rate of fluoride uptake by the bone of younger animals, and not to differences in urinary excretion. The results appear to be explained by the fact that the crystallites in developing bone are loosely organized and not compacted as in mature bone; thus, providing a much greater surface area for the rapid uptake of fluoride.

Acid-Base Status
There is a considerable body of evidence showing that the rates of fluoride absorption from the GI tract and excretion in the urine, as well as the

distribution of fluoride across the membranes of individual cells, are all dependent on pH gradients [21]. These observations are best explained by the fact that wherever there is a pH gradient across an epithelium or cell membrane separating two adjacent fluid compartments, there will also be a difference in the HF concentrations. In cases of distribution across cell membranes, HF, a highly diffusible and permeating molecule, will rapidly diffuse down its concentration gradient until the HF concentrations in the extracellular and intracellular compartments are equal. The result is that the concentration of ionic fluoride will be higher in the more alkaline compartment which, in nearly all tissues, is the extracellular fluid.

The magnitude of the pH gradient across cell membranes can be increased by alkalinizing the extracellular fluids which can be done, for example, by hyperventilating or the administration of sodium bicarbonate or sodium lactate. These actions increase extracellular pH more than intracellular pH [36]. Consequently the extracellular concentration of HF falls to a greater extent than that in the intracellular compartment which causes HF to diffuse from cells into the extracellular fluids. Thus the intracellular concentration of fluoride is reduced, thereby lowering the effects of fluoride on intracellular enzymes and transport systems. Further, the rate of fluoride absorption from the GI tract is inversely related to the pH of the stomach contents while the rate of urinary fluoride excretion is directly related to the pH of the renal tubular fluid. For all these reasons, it would be expected that the acute toxic effects of fluoride would be reduced by increasing the pH of the extracellular fluids and urine.

This expectation was confirmed in two studies with rats. The effects of pre-existing acid-base disturbances on acute fluoride toxicity [37] and of alkalosis imposed during the development of acute fluoride toxicity [38] were tested. In each study, fluoride was infused intravenously until death occurred. In the former study, acidosis or alkalosis was established before fluoride exposure by the oral administration of ammonium chloride or sodium bicarbonate, respectively. In the latter study alkalosis was established by the intravenous infusion of sodium bicarbonate with or without acetazolamide (Diamox®) during fluoride exposure. In each study, the alkalotic animals tolerated significantly higher fluoride doses and survived twice as long while the fluoride infusions continued. They also maintained higher blood pressures, heart rates, glomerular filtration rates and renal clearances of fluoride at any given plasma fluoride concentration. They died with significantly higher plasma fluoride concentrations and lower tissue-to-plasma fluoride concentration ratios. It was concluded that metabolic alkalosis, whether present before fluoride exposure or imposed during the development of toxicity, favorably influenced the clinical course and that establishing an alkalosis and a more alkaline urinary pH should be added to the therapeutic regimen.

Conclusion

As used in this chapter, acute toxicity means adverse effects that occur within a short period of time following the oral administration or ingestion of a single dose of fluoride or multiple doses within a few hours. The stomach – where the effects range from some degree of nausea to abdominal pain, bloody vomitus and diarrhea – is the first organ affected, with those latter effects signaling impending systemic effects that should be regarded as potentially fatal. Serious systemic toxic effects may occur when the amount ingested reaches the PTD of 5.0 mg/kg. It is difficult, however, to know the exact amount that was ingested, so estimations about the degree of toxicity and judgments about what actions and treatments should be taken typically depend on the early clinical signs and symptoms.

Today the most common sources of significant amounts of ingested fluoride available to most persons are fluoride-containing dental products (tables 2, 3). Compared to the frequency with which these dental products are used, cases of acute toxicity are exceedingly rare. In the US, for example, it can be reasonably estimated that fluoride dentifrices are used at least once each day by at least 200 million people and that fluoride mouth rinses, dietary supplements and professionally applied topical products are used thousands of times each day. By comparison, only about 700 persons, most of whom were young children, were treated in a healthcare facility each year from 2000 through 2003, and fewer than 100 experienced moderate or major health outcomes (table 2). The use of child-resistant caps on fluoride mouth rinse bottles and most fluoride supplement containers, warning labels on toothpaste boxes and tubes, and rational recommendations for the safe use of professionally applied products have contributed to the safe use of these products.

Nevertheless, ingestion of excessive amounts of fluoride, whether accidental or intentional, does occur and moderate and major health outcomes can follow. It is for this reason that healthcare personnel, as well as parents, should be familiar with the characteristics of acute fluoride toxicity, the sources of potentially toxic doses and how to limit access to them especially by children, the amounts of ingested fluoride that can cause harmful effects, and what to do in case of overexposures. This chapter is a source of such information.

References

1 Hodge HC, Smith FA: Biological effects of inorganic fluorides; in Simons JH (ed): Fluorine Chemistry. New York, Academic Press, 1965, pp 1–364.

2 Lidbeck WL, Hill IB, Beeman JA: Acute sodium fluoride poisoning. J Am Med Assoc 1943;121:826–827.

3 McIvor M, Baltazar RF, Beltran J, Mower MM, Wenk R, Lustgarten J, Salomon J: Hyperkalemia and cardiac arrest from fluoride exposure during hemodialysis. Am J Cardiol 1983;51:901–902.

4 McIvor ME, Cummings CC, Mower MM, Baltazar RF, Wenk RE, Lustgarten JA, Salomon J: The manipulation of potassium efflux during fluoride intoxication: implications for therapy. Toxicology 1985;37:233–239.

5 Dreisbach RH: Handbook of Poisoning. Los Altos, Langer, 1980, pp 210–213.

6 Church LI: Fluorides: use with caution. Maryland Dent Assoc J 1976;19:106.

7 Eichler HG, Lenz K, Fuhrmann M, Hruby K: Accidental ingestion of NaF tablets by children: report of a poison control center and one case. Int J Clin Pharmacol Ther Toxicol 1982;20: 334–338.

8 Dukes MNG: Side Effects of Drugs. Oxford, Excerpta Medica, 1980, p 354.

9 Bayless JM, Tinanoff N: Diagnosis and treatment of acute fluoride toxicity. J Am Dent Assoc 1985;110:209–211.

10 Gessner BD, Beller M, Middaugh JP, Whitford GM: Acute fluoride poisoning from a public water system. New Engl J Med 1994;330:95–99.

11 Watson WA, Litovitz TL, Klein-Schwartz W, Rodgers GC, Youniss J, Reid N, Rouse WG, Rembert RS, Borys D: 2003 annual report of the American Association of Poison Control Centers Toxic Exposure Surveillance System. Am J Emerg Med 2004;22:335–404.

12 US Consumer Product Safety Commission: Press release (March 26, 2009). 2009. www.cpsc.gov/Trans/ppw02.html.

13 Burrell KH, Chan JT: Systemic and topical fluorides; in Ciancio SG (ed): ADA Guide to Dental Therapeutics, ed 2. Chicago, ADA, 2000, p 233.

14 Whitford GM, Adair SM, McKnight Hanes CM, Perdue EC, Russell CM: Enamel uptake and patient exposure to fluoride: comparison of APF gel and foam. Pediat Dent 1995;17:199–203.

15 Whitford GM, Pashley DH, Garman RH: Effects of fluoride on structure and function of canine gastric mucosa. Dig Dis Sci 1997;42:2146–2155.

16 Spak C-J, Sjostedt S, Eleborg L, Veress B, Perbeck L, Ekstrand J: Studies of human gastric mucosa after application of 0.42% fluoride gel. J Dent Res 1990;69: 426–429.

17 Whitford GM, Pashley DH. Fluoride absorption: the influence of gastric acidity. Calc Tiss Int 1984;36:302–307.

18 Easmann RP, Steflik DE, Pashley DH, McKinney RV, Whitford GM: Surface changes in rat gastric mucosa induced by sodium fluoride: a scanning electron microscopic study. J Oral Pathol 1984;13: 255–264.

19 Easmann RP, Pashley DH, Birdsong NL, McKinney RV, Whitford GM: Recovery of rat gastric mucosa following single fluoride dosing. J Oral Pathol 1985;14: 779–792.

20 Pashley DH, Allison NB, Easmann R, McKinney RV, Horner JA, Whitford GM: The effects of fluoride on the gastric mucosa of the rat. J Oral Pathol 1984;13: 535–545.

21 Whitford GM: Gastric toxicity; in Myers HM (ed): The Metabolism and Toxicity of Fluoride. Basel, Karger, 1996.

22 Moody FG, Durbin RP: Effects of glycine and other instillates on concentration of gastric acid. Am J Physiol 1965;209: 122–126.

23 Ericsson Y: Monofluorophosphate physiology: general considerations. Caries Res 1983;17(suppl 1):46–55.

24 Whitford GM, Pashley DH, Allison NB: Monofluorophosphate physiology: discussion. Caries Res 1983;17(suppl 1): 69–76.

25 Müller P, Schmid K, Warnecke G, Setnikar I, Simon B: Sodium fluoride-induced gastric mucosal lesions: comparison with sodium monofluorophosphate. Gastroenterology 1992;30:252–254.

26 Shourie KL, Hein JW, Hodge HC: Preliminary studies on the caries inhibiting potential and acute toxicity of sodium monofluorophosphate. J Dent Res 1950;29:529–533.

27 Lim JK, Renaldo GJ, Chapman P: LD_{50} of SnF_2, NaF and Na_2PO_3F in the mouse compared to the rat. Caries Res 1978;12:177–179.

28 Gruninger SE, Clayton R, Chang SB, Siew C: Acute oral toxicity of dentifrice fluorides in rats and mice. J Dent Res 1988;67(special issue, abstr 1769):334.

29 Whitford GM, Birdsong-Whitford NL, Finidori C: Acute toxicity of sodium fluoride and monofluorophosphate separately or in combination in rats. Caries Res 1990;24:121–126.

30 Maynard EA, Downs WL, LeSher MF: University of Rochester Atomic Energy Project Quarterly Technical Report. UP-164, 1951, p 73–77.

31 Mornstad H: Acute sodium fluoride toxicity in rats in relation to age and sex. Acta Pharmacol Toxicol 1975;37:425–428.

32 Miller RF, Phillips PH: The enhancement of the toxicity of sodium fluoride in the rat by high dietary fat. J Nutr 1955;56:447–454.

33 Zipkin I, McClure FJ: Deposition of fluorine in the bones and teeth of the growing rat. J Nutr 1952;47:611–620.

34 Suttie JW, Phillips PH: The effect of age on fluorine deposition in the femur of the rat. Arch Biochem 1959;83:355–359.

35 Weidmann SM, Weatherell JA: The uptake and distribution of fluorine in bones. J Path Bact 1959;78:243–255.

36 Boron WF, Roos A: Comparison of microelectrode, DMO and methylamine methods of measuring intracellular pH. Am J Physiol 1976;231:799–809.

37 Reynolds KE, Whitford GM, Pashley DH: Acute fluoride toxicity: the influence of acid-base status. Toxicol Appl Pharmacol 1978;45:415–427.

38 Whitford GM, Reynolds KE, Pashley DH: Acute fluoride toxicity: influence of metabolic alkalosis. Toxicol Appl Pharmacol 1979;50:31–39.

Gary Milton Whitford, PhD, DMD
Department of Oral Biology, School of Dentistry, Medical College of Georgia
1120 15th Street
Augusta, GA 30912 (USA)
Tel. +1 706 721 0388, E-Mail gwhitfor@mcg.edu

Buzalaf MAR (ed): Fluoride and the Oral Environment.
Monogr Oral Sci. Basel, Karger, 2011, vol 22, pp 81–96

Chronic Fluoride Toxicity: Dental Fluorosis

Pamela DenBesten · Wu Li

Department of Orofacial Sciences, School of Dentistry, University of California, San Francisco, Calif., USA

Abstract

Dental fluorosis occurs as a result of excess fluoride ingestion during tooth formation. Enamel fluorosis and primary dentin fluorosis can only occur when teeth are forming, and therefore fluoride exposure (as it relates to dental fluorosis) occurs during childhood. In the permanent dentition, this would begin with the lower incisors, which complete mineralization at approximately 2–3 years of age, and end after mineralization of the third molars. The white opaque appearance of fluorosed enamel is caused by a hypomineralized enamel subsurface. With more severe dental fluorosis, pitting and a loss of the enamel surface occurs, leading to secondary staining (appearing as a brown color). Many of the changes caused by fluoride are related to cell/matrix interactions as the teeth are forming. At the early maturation stage, the relative quantity of amelogenin protein is increased in fluorosed enamel in a dose-related manner. This appears to result from a delay in the removal of amelogenins as the enamel matures. In vitro, when fluoride is incorporated into the mineral, more protein binds to the forming mineral, and protein removal by proteinases is delayed. This suggests that altered protein/mineral interactions are in part responsible for retention of amelogenins and the resultant hypomineralization that occurs in fluorosed enamel. Fluoride also appears to enhance mineral precipitation in forming teeth, resulting in hypermineralized bands of enamel, which are then followed by hypomineralized bands. Enhanced mineral precipitation with local increases in matrix acidity may affect maturation stage ameloblast modulation, potentially explaining the dose-related decrease in cycles of ameloblast modulation from ruffle-ended to smooth-ended cells that occur with fluoride exposure in rodents. Specific cellular effects of fluoride have been implicated, but more research is needed to determine which of these changes are relevant to the formation of fluorosed teeth. As further studies are done, we will better understand the mechanisms responsible for dental fluorosis.

Excess fluoride ingestion results in dental fluorosis. The mechanisms affected by long-term chronic exposure to low levels of fluoride are likely to differ from those affected by acute exposures to high levels of fluoride [1–3]. Some mechanisms affected by lower chronic fluoride levels, resulting in enamel fluorosis, are likely to be specific to this uniquely mineralizing tissue, while others may also affect other cells and tissues.

Enamel fluorosis refers to fluoride-related alterations in enamel, which occur during enamel development. These alterations become more severe with increasing fluoride intake, and time of exposure. The severity of fluorosis is related to the concentration of fluoride in the plasma,

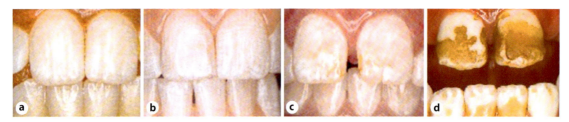

Fig. 1. Dental fluorosis. **a** Mild with slight accentuation of the perikymata. **b** Moderate, showing a white opaque appearance. **c** Moderate, white opaque enamel with some discoloration and pitting. **d** Severe.

considered to be in equilibrium with the tissue fluid that bathes the enamel organ [4, 5]. Plasma fluoride levels are influenced by many factors, including total fluoride intake, type of intake (i.e. ingested vs. inhaled), renal function, rate of bone metabolism, metabolic activity, etc. [6]. In addition to these variables, genetic factors have been shown to dictate the severity of enamel fluorosis in mice [7].

In humans, plasma fluoride concentrations resulting from long-term ingestion of 1–10 ppm fluoride in the drinking water range from 1 to 10 μmol/l. Fluorotic changes can be obtained in incisors of rodents drinking water containing 25–100 ppm fluoride; these doses also elevate plasma fluoride levels to 3–10 μmol/l, similar to those found to cause fluorosis in humans. A complicating factor in assessing the exact dose, or determining the stages of enamel formation most sensitive to fluoride, is that fluoride incorporated into bone is gradually released by continuous bone remodeling [5, 8]. Levels of plasma fluoride as low as 1.5 μmol/l (resulting from fluoride release from bone) are still capable of inducing mild enamel fluorosis in the rat incisor after the initial exposure ends [4, 8].

The effects of chronic fluoride exposure have also been linked to effects on other tissues and systems [9]. However, in this chapter, we will focus primarily on the effects of fluoride on tooth development. The largest body of research has investigated the effects of fluoride on enamel formation,

with much less known about the potential effects of fluoride on dentin formation. Therefore, most of the focus will be on enamel fluorosis. The sections of this chapter comprise:
1 Clinical manifestation, treatment and prevention of dental fluorosis;
2 Etiology and prevalence of dental fluorosis;
3 Pathology, pathogenesis and mechanism of dental fluorosis.

Clinical Manifestation, Treatment and Prevention of Dental Fluorosis

Clinical Manifestations of Dental Fluorosis
Clinically, mild cases of dental fluorosis are characterized by a white opaque appearance of the enamel, caused by increased subsurface porosity (fig. 1). The earliest sign is a change in color, showing many thin white horizontal lines running across the surfaces of the teeth, with white opacities at the newly erupted incisal end. The white lines run along the 'perikymata', a term referring to transverse ridges on the surface of the tooth, which correspond to the incremental lines in the enamel known as Striae of Retzius [10, 11].

At higher levels of fluoride exposure, the white lines in the enamel become more and more defined and thicker. Some patchy cloudy areas and thick opaque bands also appear on the involved teeth. With increased dental fluorosis, the entire tooth can be chalky white and lose transparency

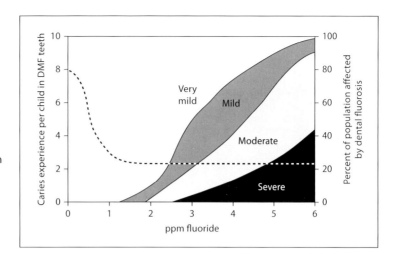

Fig. 2. Concentrations of fluoride in drinking water are related to caries incidence in children and severity of dental fluorosis. Adapted from a report of the Department of Health and Human Services of US (1991) [25].

Natural Sources of Fluoride Causing Dental Fluorosis

Dental fluorosis resulting from high fluoride levels in underground water is an issue in specific regions of the world. Fluoride can exist in an ionized form in ground waters, and in areas where the soil lacks calcium – such as occurs in areas with high levels of granite or gneiss – relatively high fluoride levels are detected in groundwater. When the level of fluoride is above 1.5 mg/l (1.5 ppm) in drinking water, dental fluorosis can occur. In some parts of Africa, China, the Middle East and southern Asia (India, Sri Lanka), as well as some areas in the Americas and Japan, high concentrations of ionic fluoride have been found in ground waters, vegetables, fruit, tea and other crops, although drinking water is usually the major source of the daily fluoride intake [23]. The atmosphere in these areas may have high levels of fluoride from dust in areas with fluoride-containing soils and gas, released from industries, underground coal fires and volcanic activities [23].

In the USA, approximately 10 million people are exposed to naturally fluoridated public water. In 1993, it was reported that 6.7 million people drank water with fluoride concentrations ≤1.2 mg/l, 1.4 million drank water with 1.3–1.9 mg/l fluoride, 1.4 million drank water with fluoride between 2.0 and 3.9 mg/l and 200,000 people ingested water with fluoride concentrations ≥4.0 mg/l [16]. Some areas have extremely high concentrations of fluoride in drinking water – such as in Colorado (11.2 mg/l), Oklahoma (12.0 mg/l), New Mexico (13.0 mg/l) and Idaho (15.9 mg/l) [9] – though water with levels higher that those recommended by the USEPA are monitored and are not used for human consumption.

Additional Sources of Fluoride Associated with Dental Fluorosis

Two primary sources have been identified as being potentially responsible for the prevalence of dental fluorosis: fluoride in drinking water and fluoride-containing dental products. Since 1945, fluoride has been used as a supplement in many public drinking water systems to control dental decay [24]. In 2000, approximately 162 million people (65.8% of the population served by public water systems) received water that contained fluoride ranging from 0.7 to 1.2 mg/l (usually 1 mg/l), depending on the local climate. The level of fluoridation is lower in high-temperature areas as people usually drink more water. The fluoridation of public drinking

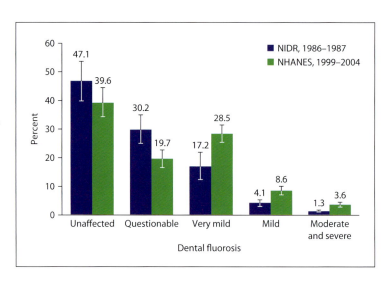

Fig. 3. Change in dental fluorosis prevalence among children aged 12–15 years participating in 2 national surveys in the USA (1986–1987 and 1999–2004). Dental fluorosis (based on Dean's fluorosis index) is defined as: very mild, mild, moderate or severe. Percentages do not sum to 100 due to rounding. Error bars = 95% CI. Sources: National Health and Nutrition Examination Survey (1999–2004) [27] and National Survey of Oral Health in U.S. School Children (1986–1987) [27].

water has significantly decreased the incidence of dental decay at a relatively low cost. In the studies by Dean and colleagues completed in the 1930s, the risk of dental fluorosis at 1 ppm fluoride in drinking water was extremely low, particularly in relation to the impact of fluoride on dental caries (fig. 2) [25]. Following these studies, water fluoridation was considered by the US Centers for Disease Control to be 1 of the 10 great public health achievements in the 20th century [26].

However, as fluoride has become more widely used in dental products (toothpastes, mouth rinses, fluoride supplements) and been incorporated into food sources (via fluoridated water), multiple sources of fluoride exposure are now related to the reported increase in the incidence of dental fluorosis. Even a small 'pea-sized' amount of toothpaste containing 1,450 ppm fluoride, would contain approximately 0.36–0.72 mg fluoride, which if consumed twice a day could contribute to fluoride levels that would increase the risk of dental fluorosis in children [21]. In the USA, the prevalence of dental fluorosis appears to be increasing. In children aged 15–17 years, the 1999–2004 National Health and Nutrition Examination Survey (NHANES) found 40.6% had very mild

or greater enamel fluorosis, up from 22.6% in the 1986–1987 study (fig. 3) [27].

The incidence of very mild and greater fluorosis in persons aged 6–39 years was 19.79% in white non-Hispanics, 32.88% in black non-Hispanics, and 25.8% in Hispanics (table 3). The increased prevalence of fluorosis in black non-Hispanics may suggest a genetic influence on fluorosis susceptibility.

Pathology, Pathogenesis and Mechanism of Dental Fluorosis

The primary pathological finding of fluorosed enamel is a subsurface porosity, along with hyper- and hypomineralized bands within the forming enamel (fig. 4) [28–34]. Fluoride can also result in mineralization-related effects on dentin formation.

Severely fluorosed human dentin is characterized by a highly mineralized sclerotic background pattern, scattered with hypomineralized porous lesions primarily in the subsurface area. Scanning electron microscope images show dentin tubules with an irregular distribution and narrow and disrupted lumina, rather than the

Table 3. Enamel fluorosis among persons aged 6–39 years by selected characteristics

	Unaffected		Questionable		Very mild		Mild		Moderate/ Severe	
	%	SE	%	SE	%	SE	%	SE	%	SE
Age group (years)										
6–11	59.81	4.07	11.80	2.50	19.85	2.12	5.83	0.73	2.71	0.59
12–15	51.46	3.51	11.96	1.84	25.33	1.98	7.68	0.93	3.56	0.59
16–19	58.32	3.30	10.21	1.70	20.79	1.78	6.65	0.67	4.03	0.77
20–39	74.86	2.28	8.83	1.23	11.15	1.22	3.34	0.58	1.81	0.39
Sex										
Male	67.65	2.63	9.99	1.45	15.65	1.52	4.58	0.54	2.12	0.39
Female	66.97	2.84	9.83	1.34	15.58	1.36	4.84	0.61	2.78	0.49
Race/ethnicity[1]										
White, non-Hispanic	69.69	3.13	10.43	1.62	14.09	1.56	3.87	0.60	1.92	0.48
Black, non-Hispanic	56.72	3.30	10.40	2.16	21.21	2.16	8.24	0.82	3.43	0.54
Mexican-American	65.25	3.89	8.95	1.29	15.93	2.24	5.05	0.72	4.82[2]	1.81
Poverty Status[3]										
<100% FPL	68.02	3.21	10.67	1.64	14.28	1.73	4.07	0.69	2.97	0.66
100–199% FPL	66.92	2.91	9.11	1.79	16.11	1.46	5.21	0.78	2.65	0.56
≥200% FPL	66.88	2.75	10.73	1.33	15.56	1.56	4.83	0.50	2.00	0.37
Total	67.40	2.65	9.91	1.35	15.55	1.37	4.69	0.49	2.45	0.40

Data from National Health and Nutrition Examination Survey (1999–2002) [27] and calculated using Dean's index. All estimates are adjusted by age (single years) and sex to the USA 2000 standard population, except sex, which is adjusted only by age.
[1] Calculated using 'other race/ethnicity' and 'other Hispanic' in the denominator.
[2] Unreliable estimate: the standard error is 30% the value of the point estimate, or greater.
[3] Percentage of the federal poverty level (FPL), which varies by income and number of persons living in the household.

regular-appearing lumina seen in normal dentin [35].

The pathogenesis of dental fluorosis is related to physiological conditions, including body weight, rate of skeletal growth and remodeling, nutrition, and renal function [36–38]. Bone is a reservoir of fluoride, as fluoride is incorporated in the forming apatite crystals, and this ion can also be released from these crystals as bone remodels. Therefore, rapid bone growth, as occurs in the growing child, will remove fluoride from the blood stream, possibly reducing the risk of dental fluorosis by lowering serum fluoride levels [8, 39]. Nutrition is also important for controlling the serum level of fluoride, as ions such as calcium, magnesium and aluminum can reduce the bioavailability of fluoride. A deficiency in these ions in food can also affect (enhance) fluoride uptake [40].

Genetic background appears to have role in the pathogenesis of dental fluorosis. This may be the reason why in human populations, individuals

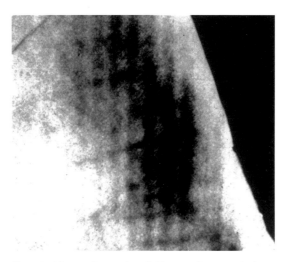

Fig. 4. Microradiograph of fluorosed enamel from Colorado Springs. Note the radiolucent outer third of the enamel with a well-calcified surface layer. From Newbrun [96], reprinted with permission.

drinking water with similar fluoride contents have a wide range of severity of dental fluorosis (fig. 2). Evidence for a genetic component to fluoride susceptibility comes from work by Everett et al. [7], which tested 12 different in-bred mouse strains to compare their susceptibility to fluoride. Mouse teeth have been found to be an excellent model for human tooth formation, and in Everett's study, they found that some mouse strains were highly susceptible to fluoride-related dental fluorosis, while other strains were highly fluorosis resistant. They concluded that there is a genetic component to dental fluorosis susceptibility [41].

Stages of Tooth Formation and Stage-Specific Effects of Chronic Fluoride Exposure
Fluoride is a single highly electronegative ion that interacts with the cells and matrix at the different stages of enamel formation in relation to fluoride dose and time of exposure. Tooth enamel development can be divided into 4 major stages: pre-secretory, secretory, transition and maturation stages, all with unique properties that affect

fluoride susceptibility. Most of the studies of the mechanisms of fluoride in forming fluorosed enamel have used the rodent incisor or molars as a model, as it is not possible to do similar studies using human teeth. The rodent incisor is a continuously erupting tooth, with all stages of enamel formation present in each tooth, whereas the molar is a rooted tooth, which begins formation in utero. As previously mentioned, though rodents require the ingestion of much higher levels of fluoride in the drinking water (10–20 times) as compared to humans, the serum levels at which fluorosis is formed in rodents and humans is similar.

Pre-secretory ameloblasts differentiate into secretory ameloblasts after the dentin matrix begins to mineralize. The pre-secretory ameloblasts and overlying cells of the enamel organ, including the enamel knot, are thought to influence the tooth morphogenesis. However, there is no evidence that exposure of developing teeth to physiological levels of fluoride in vivo [42] and in organ culture [43–46] affects tooth morphogenesis. Even in teeth with severe fluorosis, the size and form of the teeth are not changed [47].

As the pre-ameloblasts differentiate to secretory ameloblasts, they begin to secrete enamel matrix proteins, and lay down a thin layer of aprismatic enamel deposited against mantle dentin. As the secretory ameloblast Tomes' processes form, the inner enamel layer, which constitutes the bulk of enamel, begins to be laid down. This enamel matrix consists of prismatic enamel with rod (or prisms) and interrod structures (interprismatic enamel) formed by the Tomes' processes of fully differentiated secretory ameloblasts. These cells secrete matrix protein (predominantly amelogenins) into the enamel space through which thin but long enamel crystals grow preferentially in length in the wake of the retreating cells.

Secretory stage ameloblasts exposed to high chronic levels of fluoride have a somewhat disrupted morphology and increased numbers of vacuoles at the apical border. Chronic exposure to fluoride in drinking water or repeated injections of moderate fluoride doses reduces the thickness of enamel by about 10% [42, 48]. Although this suggests that chronic exposure to fluoride reduces biosynthesis of matrix by secretory ameloblasts, there is no evidence to support this [1, 49, 50]. Instead, the small reduction in enamel thickness may be attributed to a limited disruption of vesicular transport in fluorotic secretory ameloblasts and subsequent intracellular degradation of a minor portion of the matrix by the lysosomal system [51–53]. Alternatively, the reduction in enamel thickness may be related to an effect of fluoride on crystal elongation in the secretory stage.

At the end of secretion, the ameloblasts lose their Tomes' process and deposit a final layer of aprismatic enamel with small crystals. The cells transform via a short transitional stage, where enamel matrix proteins undergo rapid proteolysis, leaving the porous enamel matrix characteristic of this transition stage.

Late secretory-transitional cell stage ameloblasts appear to be more sensitive to fluoride than early and fully secretory ameloblasts. In hamster molar tooth germs, a dose of 4.5 mg/kg fluoride induces the late secretory to transitional cells, but not early secretory ameloblasts to detach occasionally from the surface and form subameloblastic cysts. The enamel below the cysts under late secretory ameloblasts will give rise to the shallow occlusal pits, often seen in severely fluorosed teeth in various species [47, 54–60]. This stage of development is likely also to be associated with the formation of accentuated perikymata that is clinically the first sign of enamel fluorosis.

In the maturation stage, the ameloblasts modulate cyclically from cells with a smooth-ended to a ruffle-ended distal membrane, the latter with characteristics of resorbing cells. During this modulation, matrix proteins continue to be removed from the extracellular space, and mineralization increases to form a fully mineralized enamel matrix. Amelogenin proteins are retained in the fluorosed rat enamel matrix at this stage of enamel formation [50, 61].

Maturation ameloblasts of adult rat incisors [42] are shorter, and fluorotic enamel organs have a disrupted maturation ameloblast modulation [42, 62, 63]. The first modulation bands that disappear during fluoride exposure are the most incisal smooth-ended ameloblasts. At prolonged exposure other smooth-ended bands disappear one by one in an incisal to apical direction [62]. In addition to changes in modulation, fluoride also reduces the cyclic uptake of ^{45}Ca labeling in a similar pattern [62]. When fluoride exposure is discontinued, smooth-ended bands reappear starting from the youngest most apical part towards older more incisal bands. This suggests that the fluoride effects on ameloblast modulation are reversible, and that the young modulating cells recover more rapidly than older ameloblasts. After eruption, the enamel is exposed to mineral ions of the oral fluids, including fluoride, which can influence the composition of the outer layers of enamel.

Direct Effects of Fluoride on Ameloblasts
Ameloblasts and tooth organs exposed to high (millimolar) levels of fluoride in vitro, which would be much greater than the micromolar levels of fluoride found in the plasma carrying fluoride ions to tooth organs in vivo, show many alterations. These include changes in the structure of early secretory ameloblasts, reduced protein synthesis, altered cell proliferation, apoptosis, stress-related protein upregulation and elevation of F-actin [64–67]. However, some of these same changes are not readily apparent in vivo, and therefore, the effects of fluoride when examined in culture, must be carefully analyzed for biological relevance.

However, there are in vitro data indicating that ameloblasts can be sensitive to low levels of fluoride. Human primary enamel organ epithelial cells grown in culture show that exposure to fluoride levels as low as 5 μmol/l results in reduced expression of the secretory stage matrix metalloproteinase 20 (MMP-20) [68], mediated by JNK/c-Jun signaling [69]. These results suggest that fluoride may have specific effects on ameloblast

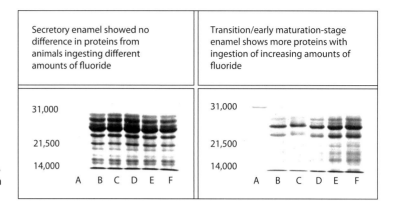

Fig. 5. SDS PAGE separation of proteins in secretory and maturation stages of enamel matrix of fluoride-treated and untreated rat tooth. A = Standard; B = 0 ppm; C = 10 ppm; D = 25 ppm; E = 50 ppm; F = 100 ppm. From DenBesten [50], reprinted with permission.

differentiation mediated through MAP-kinase signaling.

Rodent studies have shown that ingestion of fluoride alters the number of bands of smooth ended ameloblasts and their rate of modulation in the maturation stage ameloblasts [42, 62]. However, there is currently no evidence to determine whether these changes in maturation stage ameloblast modulation are a direct effect of fluoride, or more likely, in response to matrix-mediated alterations related to fluoride exposure to the developing enamel matrix.

At extremely high levels of ingested fluoride (150 ppm) in the drinking water, ameloblasts have been shown to exhibit apoptosis and endoplasmic reticulum stress responses [65]; however, at lower levels (75 ppm) these effects were not noted. Further studies at lower fluoride levels will need to be done to determine whether this is a potential mechanism relevant to chronic fluoride toxicity in humans.

Fluoride-Related Alterations of the Forming Enamel Matrix May Indirectly Affect Ameloblast Function

The extracellular enamel matrix proteins include amelogenins, ameloblastin and enamelin, all of which support and modulate enamel crystal formation [70]. Amelogenin is the chief structural protein constituting 90–95% of total proteins in the enamel protein matrix [71]. Amelogenin and the other matrix proteins are hydrolyzed by matrix proteinases as enamel forms, allowing replacement of the protein matrix with an organized hydroxyapatite structure. MMP-20 is the proteinase primarily responsible for the initial hydrolysis of amelogenins in the secretory enamel matrix, while kallikrein 4 (KLK4) is the predominant proteinase in the transition/maturation stage [72, 73].

An analysis of proteolytic activity in enamel matrix, isolated from secretory and maturation stage rat enamel, showed a significantly reduced activity in early maturation stage enamel isolated from rats ingesting 100 ppm fluoride (5–10 μM serum fluoride), as compared to control maturation enamel [74]. This effect of fluoride ingestion in decreasing matrix proteinase activity correlates to an increased retention of amelogenin proteins in maturation stage fluorosed enamel in a dose-dependent manner (fig. 5). Matrix proteins disappear from nonfluorosed enamel in the maturation stage, but are retained in fluorosed enamel, with increased retention at higher levels of ingested fluoride [48, 50].

This retention of amelogenin proteins could delay final mineralization of the enamel matrix, contributing to subsurface hypomineralization characteristic of fluorosed enamel. The reason for this retention of amelogenins is most likely related to altered proteolytic activity in the fluorosed enamel matrix.

Reduced Proteolytic Activity May Be due to the Effects of Fluoride Incorporation into Growing Enamel Crystals

Crystals in sound enamel are long, and the dynamics of enamel crystal growth, size of the crystals and their shape are well controlled by matrix proteins during enamel formation [75–77]. Some studies report that crystals isolated from fluorosed enamel have a significantly greater diameter than crystals in sound enamel, as determined by high-resolution electron microscopy [78], X- ray diffraction of powdered enamel samples [79] or scanning microscopy of fractured inner enamel specimens [80]. Some organ culture studies have shown large flattened hexagonal crystals mixed with many small irregularly shaped crystals in hypermineralized areas [81, 82]. However, other studies reported no differences between fluorotic and normal human crystals [28, 83].

There is, however, no doubt that the fluoride content of crystals in fluorosed enamel is greater than that of normal enamel. Fluoride substitutes for hydroxyl groups in enamel carbonated hydroxyapatite crystals, altering the crystalline structures and surface characteristics. To determine whether an increased fluoride content of the apatite crystals could affect matrix/proteinase interactions, we measured the binding of recombinant human amelogenin to synthetic carbonated hydroxyapatite crystals.

The initial rate of amelogenin binding and the total amount of amelogenin bound to fluoride-containing carbonated hydroxyapatite was greater than that in the control carbonated hydroxyapatite [84]. These results suggest that fluoride incorporation into the crystal lattice alters the crystal surface to enhance amelogenin binding, potentially contributing to the increased amount of amelogenin and the inhibition of crystal growth in fluorosed enamel.

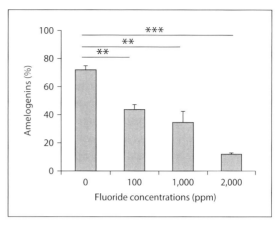

Fig. 6. Degradation of amelogenin adsorbed on apatite crystals by MMP-20. Amelogenins were pre-bound to carbonated hydroxyapatite crystals containing different amounts of fluoride (X-axis) and then degraded by MMP-20. Y-axis indicates the percentages of amelogenins degraded by MMP-20 from apatite crystals as compared to the amount of amelogenin initially bound. Note the decreased degradation of amelogenin from the apatite crystal surface as the concentration of fluoride in the apatite increases (unpublished data).

In further investigation of the role of fluoride incorporation into apatite on amelogenin processing, we characterized hydrolysis of amelogenins bound to fluoride-containing apatites by recombinant MMP-20 or KLK-4. When fluoride was in solution, amelogenin hydrolysis by MMP-20 was reduced only at 1,000 ppm (52 mM, which is far higher than physiological levels of fluoride in enamel fluids). However, incorporation of fluoride into apatite significantly delayed MMP-20 hydrolysis of the adsorbed amelogenin in a dose-dependent manner (fig. 6) even at the lowest level of fluoride-containing apatite (100 ppm F). This same effect of reduced amelogenin hydrolysis was found when amelogenins were hydrolyzed from fluoride-containing apatites with recombinant KLK-4 (unpublished results).

The levels of fluoride incorporated into the apatite crystals in these in vitro studies are biologically relevant. Although the enamel fluid surrounding the ameloblasts is likely to contain no more than 10 μM (0.19 ppm) fluoride, fluoride is incorporated into the growing crystals in concentrations ranging from 10 ppm near the dental-enamel junction to several thousand ppm at the enamel surface [85]. Fluoride-containing apatite with fluoride concentrations of 100 ppm are found in the inner enamel (300 μm from the surface) of human teeth with minimal (mild) fluorosis [85]. The higher fluoride-containing apatite (approximately 2,000 ppm F) is similar to that found in the mid-layer of enamel (150 μm from the surface) of severely fluorosed human teeth. Therefore, these studies indicate that the reduced hydrolysis of amelogenin found in fluorosed maturation stage enamel [1, 51] may be due to the reduction in the rate of hydrolysis of amelogenins bound to fluoride-containing enamel crystals.

These effects of fluoride incorporation on hydrolysis of apatite-bound amelogenins is consistent with the observation that fluoride-induced subsurface hypomineralization can independently occur in the maturation stage only [59, 62, 86]. Mineralization defects in fluorosed rat incisor maturation stage enamel are characterized by the development of a generalized hypomineralized porous subsurface area along the entire crown enamel [4, 87–90]. This type of defect correlates to the porous white opacities seen clinically.

Potential Effects of Matrix pH on Fluoride-Related Changes in Enamel Formation
Matrix protein removal may also be influenced by fluoride-mediated changes in pH during apatite crystal formation. Formation of apatite results in the formation of a substantial number of protons [$10Ca^{2+} + 6\ HPO_4^{2-} + 2H_2O \rightarrow Ca_{10}(PO_4)_6(OH)_2 + 8H^+$] that need to be neutralized. Amelogenins bind as many as 12 protons per molecule [91]. However, if this amelogenin buffering system is either not available, or is saturated, it is conceivable that a fluoride-induced pH drop could alter the amelogenin tertiary structure and affect its function [92].

Abundant amelogenins generated by secretory ameloblasts may be a potent contributor to controlling pH at the secretory stage, where the pH is maintained at neutral [76, 93]. At the end of the secretory stage, enamel matrix proteinases are activated, and at the transition stage, enamel matrix proteins are rapidly lost. At this stage, the cell junctions between the ameloblasts are open, allowing fluoride to readily move from the serum to the enamel matrix. The presence of increased amounts of fluoride in the transition stage may make this stage highly susceptible to the effects of fluoride on enamel formation.

At the maturation stage, the pH in the enamel matrix changes periodically between acidic (pH 5.8) and neutral (pH 7.2) as ameloblasts modulate [94, 95]. If we assume that the acidification of the enamel matrix has a role in ameloblast modulation from ruffle-ended to smooth-ended ameloblasts, in dental fluorosis, changes in matrix pH secondary to fluoride-enhanced mineral deposition could contribute to the delay in the transition from ruffle-ended to smooth-ended ameloblasts. This delay in ameloblast modulation (which is a characteristic of fluorosed maturation ameloblasts) could possibly contribute to the delay in removal of amelogenins which occur in fluorosed enamel.

Particularly at this final stage of enamel mineralization, Bronckers et al. [93] have hypothesized that fluoride in the enamel matrix may enhance mineralization resulting in localized hypermineralization, requiring the ameloblasts to pump additional bicarbonate into the extracellular enamel matrix. This hypermineralization would deplete the local reservoir or free calcium ions, resulting in a subsequent band of hypomineralized enamel. This hypothesis is supported by a recent study showing an upregulation of mRNA for the pH regulator NBCe1 in fluorosed maturation stage ameloblasts as compared to control maturation ameloblasts [92].

In summary, the mechanisms by which fluoride alters enamel maturation are multi-factorial. We propose a multi-stage model for the formation of fluorosed enamel, as follows:

(1) Crystals forming in the secretory stage of enamel have an increased fluoride content and therefore bind more amelogenin.

(2) Hydrolysis of amelogenins by proteinases is delayed by altered amelogenin interactions with the fluoride-containing hydroxyapatite crystals.

(3) At the transition stage, fluoride is rapidly deposited into the porous enamel matrix between the open cell junctions, resulting in increased formation of fluoride-containing apatite, and a delay in protein hydrolysis secondary to altered mineral/matrix interactions.

(4) The net result of these fluoride-related effects in the secretory and transition stages is retention of amelogenins in the maturation stage. This delay in removal of amelogenins increases the relative pH in the maturation stage under ruffle-ended ameloblasts as amelogenins buffer the increased protons resulting from mineral formation.

(5) The reduced acidification of the matrix under ruffle-ended ameloblasts further delays modulation to smooth-ended ameloblasts, resulting in fewer bands of modulating ameloblasts.

(6) In late maturation, when amelogenins are finally removed (or in mild dental fluorosis with minimal amelogenin retention), fluoride-mediated hypermineralization may increase the local acidification affecting ameloblast function, such as ion transport activities. Although porous subsurface enamel is the major phenotype of fluorosed enamel, successive layers of hypomineralized and hypermineralized enamel are also a characteristic of the fluorosed enamel matrix [4, 89].

It is likely that there are additional effects of fluoride, including other indirect effects on cells at different stages of formation, and that in the course of our and others' studies this model and our understanding of the mechanisms (including more potential direct cellular effects) will be expanded.

Acknowledgements

The authors would thank Antonius Bronkers and Donacian Lyaruu from ACTA University, Amsterdam for their insightful discussions, as well as funding support from NIH grant # R01DE013508.

References

1 Aoba T, Moreno EC, Tanabe T, Fukae M: Effects of fluoride on matrix proteins and their properties in rat secretory enamel. J Dent Res 1990;69:1248–1250.

2 Richards A: Nature and mechanisms of dental fluorosis in animals. J Dent Res 1990;69(spec No):701–705, discussion 721.

3 Giambro NJ, Prostak K, DenBesten PK: Characterization of fluorosed human enamel by color reflectance, ultrastructure, and elemental composition. Caries Res 1995;29:251–257.

4 Angmar-Månsson B, Ericsson Y, Ekberg O: Plasma fluoride and enamel fluorosis. Calcif Tissue Res 1976;22:77–84.

5 Angmar-Månsson B, Whitford GM: Enamel fluorosis related to plasma F levels in the rat. Caries Res 1984;18: 25–32.

6 Angmar-Månsson B, Lindh U, Whitford GM: Enamel and dentin fluoride levels and fluorosis following single fluoride doses: a nuclear microprobe study. Caries Res 1990;24:258–262.

7 Everett ET, McHenry MA, Reynolds N, et al: Dental fluorosis: variability among different inbred mouse strains. J Dent Res 2002;81:794–798.

8 Angmar-Månsson B, Whitford GM: Environmental and physiological factors affecting dental fluorosis. J Dent Res 1990:706–713.

9 National Research Council: Fluoride in Drinking Water: A Scientific Review of EPA's Standards. Washington, National Academies Press, 2006, p 530.

10 Moller IJ: Fluorides and dental fluorosis. Int Dent J 1982;32:135–147.

11 Kroncke A: Perikymata (in German). Dtsch Zahnarztl Z 1966;21:1397–1401.

12 Smith GE: Fluoride, teeth and bone. Med J Aust 1985;143:283–286.

13 McKay FS: The study of mottled enamel (dental fluorosis). J Am Dent Assoc 1952;44:133–137.

14 Mottled enamel. Am J Public Health Nations Health 1933;23:47–48.

15 Ockerse T, Wasserstein B: Stain in mottled enamel. J Am Dent Assoc 1955;50:536–538.

16 Subcommittee on Health Effects of Ingested Fluoride (National Research Council): Health Effects of Ingested Fluoride. Washington, National Academy of Sciences, 1993;169.

17 Dean HT: Fluorine in the control of dental caries. J Am Dent Assoc 1956;52:1–8.

18 Dean HT: Endemic fluorosis and its relation to dental caries: 1938. Public Health Rep 2006;121(suppl 1):213–219, discussion 212.

19 Thylstrup A, Fejerskov O, Mosha HJ: A polarized light and microradiographic study of enamel in human primary teeth from a high fluoride area. Arch Oral Biol 1978;23:373–380.

20 USEPA: Fluorine (soluble fluoride) (CASRN 7782-41-4). Washington, Integrated Risk Information System, 1987.

21 Institute of Medicine: Dietary reference intakes: for calcium, phosphorus, magnesium, vitamin D, and fluoride. Washington, National Academies Press, 1997.

22 Levy SM, Guha-Chowdhury N: Total fluoride intake and implications for dietary fluoride supplementation. J Public Health Dent 1999;59:211–223.

23 World Health Organization: Water for Life: Making It Happen. Geneva, WHO, 2005.

24 Lennon MA: One in a million: the first community trial of water fluoridation. Washington, National Institutes of Health, 2006.

25 Service PH: Review of Fluoride Benefits and Risks, Report of the ad hoc Submommittee on Fluoride of the Committee to Coordinate Environmental Health and Related Programs. Washington, Department of Health and Human Services, 1991.

26 Centers for Disease Control: Ten Great Public Health Achievements in the 20th Century. Washington, CDC, 2008.

27 Beltrán-Aguilar ED, Barker LK, Canto MT, et al: Surveillance for dental caries, dental sealants, tooth retention, edentulism, and enamel fluorosis – United States, 1988–1994 and 1999–2002. MMWR Surveill Summ 2005;54:1–43.

28 Fejerskov O, Johnson NW, Silverstone LM: The ultrastructure of fluorosed human dental enamel. Scand J Dent Res 1974;82:357–372.

29 Fejerskov O, Larsen MJ, Josephsen K, Thylstrup A: Effect of long-term administration of fluoride on plasma fluoride and calcium in relation to forming enamel and dentin in rats. Scand J Dent Res 1979;87:98–104.

30 Fejerskov O, Silverstone LM, Melsen B, Moller IJ: Histological features of fluorosed human dental enamel. Caries Res 1975;9:190–210.

31 Fejerskov O, Thylstrup A, Larsen MJ: Clinical and structural features and possible pathogenic mechanisms of dental fluorosis. Scand J Dent Res 1977;85:510–534.

32 Fejerskov O, Yanagisawa T, Tohda H, et al: Posteruptive changes in human dental fluorosis – a histological and ultrastructural study. Proc Finn Dent Soc 1991;87:607–619.

33 Kidd EA, Thylstrup A, Fejerskov O: The histopathology of enamel caries in fluorosed deciduous teeth. Caries Res 1981;15:346–352.

34 Kierdorf U, Kierdorf H, Fejerskov O: Fluoride-induced developmental changes in enamel and dentine of European roe deer (*Capreolus capreolus* L.) as a result of environmental pollution. Arch Oral Biol 1993;38:1071–1081.

35 Rojas-Sanchez F, Alaminos M, Campos A, Rivera H, Sanchez-Quevedo MC: Dentin in severe fluorosis: a quantitative histochemical study. J Dent Res 2007;86:857–861.

36 Ekstrand J, Spak CJ, Ehrnebo M: Renal clearance of fluoride in a steady state condition in man: influence of urinary flow and pH changes by diet. Acta Pharmacol Toxicol (Copenh) 1982;50:321–325.

37 Jarnberg PO, Ekstrand J, Irestedt L, Santesson J: Renal fluoride excretion during and after enflurane anaesthesia: dependency on spontaneous urinary pH-variations. Acta Anaesthesiol Scand 1980;24:129–134.

38 Spak CJ, Berg U, Ekstrand J: Renal clearance of fluoride in children and adolescents. Pediatrics 1985;75:575–579.

39 Pendrys DG, Stamm JW: Relationship of total fluoride intake to beneficial effects and enamel fluorosis. J Dent Res 1990;69(spec No):529–538, discussion 556–527.

40 Taves DR: Dietary intake of fluoride ashed (total fluoride) v. unashed (inorganic fluoride) analysis of individual foods. Br J Nutr 1983;49:295–301.

41 Everett ET, Yan D, Weaver M, Liu L, Foroud T, Martinez-Mier EA: Detection of dental fluorosis-associated quantitative trait loci on mouse chromosomes 2 and 11. Cells Tissues Organs 2009;189:212–218.

42 Smith CE, Nanci A, Denbesten PK: Effects of chronic fluoride exposure on morphometric parameters defining the stages of amelogenesis and ameloblast modulation in rat incisors. Anat Rec 1993;237:243–258.

43 Bronckers AL, Jansen LL, Wöltgens JH: A histological study of the short-term effects of fluoride on enamel and dentine formation in hamster tooth-germs in organ culture in vitro. Arch Oral Biol 1984;29:803–810.

44 Bronckers AL, Jansen LL, Wöltgens JH: Long-term (8 days) effects of exposure to low concentrations of fluoride on enamel formation in hamster tooth-germs in organ culture in vitro. Arch Oral Biol 1984;29:811–819.

45 Kerley MA, Kollar EJ: Regeneration of tooth development in vitro following sodium fluoride treatment. Am J Anat 1977;149:181–195.

46 Levenson GE: The effect of fluoride on ameloblasts of mouse molar tooth germs 'in vitro'. J Biol Buccale 1980;8:255–263.

47 Kierdorf H, Kierdorf U: Disturbances of the secretory stage of amelogenesis in fluorosed deer teeth: a scanning electron-microscopic study. Cell Tissue Res 1997;289:125–135.

48 Zhou R, Zaki AE, Eisenmann DR: Morphometry and autoradiography of altered rat enamel protein processing due to chronic exposure to fluoride. Arch Oral Biol 1996;41:739–747.

49 Bronckers ALJJ, Wöltgens JHM: Short term effects of fluoride on biosynthesis of enamel-matrix proteins and dentin collagens and on mineralization during hamster tooth-germ development in organ culture. Archs Oral Biol 1985;39:181–185.

50 DenBesten PK: Effects of fluoride on protein secretion and removal during enamel development in the rat. J Dent Res 1986;65:1272–1277.

51 Bronckers AL, Lyaruu DM, Bervoets TJ, Wöltgens JH: Fluoride enhances intracellular degradation of amelogenins during secretory phase of amelogenesis of hamster teeth in organ culture. Connect Tissue Res 2002;43:456–465.

52 Matsuo S, Inai T, Kurisu K, Kiyomiya K, Kurebe M: Influence of fluoride on secretory pathway of the secretory ameloblast in rat incisor tooth germs exposed to sodium fluoride. Arch Toxicol 1996;70:420–429.

53 Monsour P, Harbrow J, Warshawsky H: Effects of acute doses of sodium fluoride on the morphology and the detectable calcium associated with secretory ameloblasts in rat incisors. J Histchem Cytochem 1989;37:463–471.

54 Fejerskov O, Larsen MJ, Richards A, Baelum V: Dental tissue effects of fluoride. Adv Dent Res 1994;8:15–31.

55 Kardos TB, Hunter AR, Hubbard MJ: Scanning electron microscopy of trypsin-treated enamel from fluorosed rat molars. Adv Dent Res 1989;3:183–187.

56 Milhaud GE, Charles E, Loubiere ML, Kolf-Clauw M, Joubert C: Effects of fluoride on secretory and postsecretory phases of enamel formation in sheep molars. Am J Vet Res 1992;53:1241–1247.

57 Nelson DG, Coote GE, Vickridge IC, Suckling G: Proton microprobe determination of fluorine profiles in the enamel and dentine of erupting incisors from sheep given low and high daily doses of fluoride. Arch Oral Biol 1989;34:419–429.

58 Richards A, Kragstrup J, Josephsen K, Fejerskov O: Dental fluorosis developed in post-secretory enamel. J Dent Res 1986;65:1406–1409.

59 Suckling G, Thurley DC, Nelson DG: The macroscopic and scanning electron-microscopic appearance and microhardness of the enamel, and the related histological changes in the enamel organ of erupting sheep incisors resulting from a prolonged low daily dose of fluoride. Arch Oral Biol 1988;33:361–373.

60 Susheela AK, Bhatnagar M: Fluoride toxicity: a biochemical and scanning electron microscopic study of enamel surface of rabbit teeth. Arch Toxicol 1993;67:573–579.

61 DenBesten PK, Crenshaw MA: The effects of chronic high fluoride levels on forming enamel in the rat. Arch Oral Biol 1984;29:675–679.

62 DenBesten PK, Crenshaw MA, Wilson MH: Changes in the fluoride-induced modulation of maturation stage ameloblasts of rats. J Dent Res 1985;64:1365–1370.

63 Nishikawa S, Josephsen K: Cyclic localization of actin and its relationship to junctional complexes in maturation ameloblasts of the rat incisor. Anat Rec 1987;219:21–31.

64 Bartlett JD, Dwyer SE, Beniash E, Skobe Z, Payne-Ferreira TL: Fluorosis: a new model and new insights. J Dent Res 2005;84:832–836.

65 Kubota K, Lee DH, Tsuchiya M, et al: Fluoride induces endoplasmic reticulum stress in ameloblasts responsible for dental enamel formation. J Biol Chem 2005;280:23194–23202.

66 Li Y, Decker S, Yuan ZA, et al: Effects of sodium fluoride on the actin cytoskeleton of murine ameloblasts. Arch Oral Biol 2005;50:681–688.

67 Yan Q, Zhang Y, Li W, Denbesten PK: Micromolar fluoride alters ameloblast lineage cells in vitro. J Dent Res 2007;86:336–340.

68 Zhang Y, Yan Q, Li W, DenBesten PK: Fluoride down-regulates the expression of matrix metalloproteinase-20 in human fetal tooth ameloblast-lineage cells in vitro. Eur J Oral Sci 2006;114(suppl 1):105–110, discussion 127–109, 380.

69 Zhang Y, Li W, Chi HS, Chen J, Denbesten PK: JNK/c-Jun signaling pathway mediates the fluoride-induced down-regulation of MMP-20 in vitro. Matrix Biol 2007;26:633–641.

70 Robinson C, Brookes SJ, Shore RC, Kirkham J: The developing enamel matrix: nature and function. Eur J Oral Sci 1998;106(suppl 1):282–291.

71 Fincham AG, Moradian-Oldak J: Recent advances in amelogenin biochemistry. Connect Tissue Res 1995;32:119–124.

72 Hu JC, Ryu OH, Chen JJ, et al: Localization of EMSP1 expression during tooth formation and cloning of mouse cDNA. J Dent Res 2000;79:70–76.

73 Hu JC, Sun X, Zhang C, et al: Enamelysin and kallikrein-4 mRNA expression in developing mouse molars. Eur J Oral Sci 2002;110:307–315.

74 DenBesten PK, Yan Y, Featherstone JD, et al: Effects of fluoride on rat dental enamel matrix proteinases. Arch Oral Biol 2002;47:763–770.

75 Moradian-Oldak J, Jimenez I, Maltby D, Fincham AG: Controlled proteolysis of amelogenins reveals exposure of both carboxy- and amino-terminal regions. Biopolymers 2001;58:606–616.

76 Simmer JP, Fincham AG: Molecular mechanisms of dental enamel formation. Crit Rev Oral Biol Med 1995;6:84–108.

77 Smith CE, Nanci A: Protein dynamics of amelogenesis. Anat Rec 1996;245:186–207.

78 Kerebel B, Daculsi G: Ultrastructural and crystallographic study of human enamel in endemic fluorosis (in French). J Biol Buccale 1976;4:143–154.

79 Vieira A, Hancock R, Dumitriu M, et al: How does fluoride affect dentin microhardness and mineralization? J Dent Res 2005;84:951–957.

80 Sundstrom B, Jongebloed WL, Arends J: Fluorosed human enamel. A SEM investigation of the anatomical surface and outer and inner regions of mildly fluorosed enamel. Caries Res 1978;12:329–338.

81 Yanagisawa T, Takuma S, Fejerskov O: Ultrastructure and composition of enamel in human dental fluorosis. Adv Dent Res 1989;3:203–210.

82 Yanagisawa T, Takuma S, Tohda H, Fejerskov O, Fearnhead RW: High resolution electron microscopy of enamel crystals in cases of human dental fluorosis. J Electron Microsc (Tokyo) 1989;38:441–448.

83 Robinson C, Yamamoto K, Connell SD, et al: The effects of fluoride on the nanostructure and surface pK of enamel crystals: an atomic force microscopy study of human and rat enamel. Eur J Oral Sci 2006;114(suppl 1):99–104, discussion 127–109, 380.

84 Tanimoto K, Le T, Zhu L, et al: Effects of fluoride on the interactions between amelogenin and apatite crystals. J Dent Res 2008;87:39–44.

85 Richards A, Fejerskov O, Baelum V: Enamel fluoride in relation to severity of human dental fluorosis. Adv Dent Res 1989;3:147–153.

86 Richards A, Kragstrup J, Nielsen-Kudsk F: Pharmacokinetics of chronic fluoride ingestion in growing pigs. J Dent Res 1985;64:425–430.

87 Angmar-Månsson B, Whitford GM: Plasma fluoride levels and enamel fluorosis in the rat. Caries Res 1982;16:334–339.

88 Kierdorf H, Kierdorf U, Richards A, Josephsen K: Fluoride-induced alterations of enamel structure: an experimental study in the miniature pig. Anat Embryol (Berl) 2004;207:463–474.

89 Richards A, Likimani S, Baelum V, Fejerskov O: Fluoride concentrations in unerupted fluorotic human enamel. Caries Res 1992;26:328–332.

90 Shinoda H: Effect of long-term administration of fluoride on physico-chemical properties of the rat incisor enamel. Calcif Tissue Res 1975;18:91–100.

91 Ryu OH, Hu CC, Simmer JP: Biochemical characterization of recombinant mouse amelogenins: protein quantitation, proton absorption, and relative affinity for enamel crystals. Connect Tissue Res 1998;38:207–214, discussion 241–206.

92 Zheng L, Zhang Y, He P, Kim J, Schneider R, Bronckers AL, Lyaruu DM, DenBesten PK: NBCe1 in Mouse and Human Ameloblasts may be Indirectly Regulated by Fluoride. J Dent Res 2011;90:782–787.

93 Bronckers AL, Lyaruu DM, DenBesten PK: The impact of fluoride on ameloblasts and the mechanisms of enamel fluorosis. J Dent Res 2009;88:877–893.

94 Sasaki S, Takagi T, Suzuki M: Cyclical changes in pH in bovine developing enamel as sequential bands. Arch Oral Biol 1991;36:227–231.

95 Smith CE: Cellular and chemical events during enamel maturation. Crit Rev Oral Biol Med 1998;9:128–161.

96 Newbrun E, Brudevold F: Studies on the physical properties of fluorosed enamel-I: Microradiography. Arch Oral Biol 1960;2:15–20.

Pamela DenBesten
Department of Orofacial Sciences, School of Dentistry, University of California, San Francisco
513 Parnassus Avenue
San Francisco, CA 94143 (USA)
Tel. +1 415 502 7828, E-Mail Pamela.DenBesten@ucsf.edu

Buzalaf MAR (ed): Fluoride and the Oral Environment.
Monogr Oral Sci. Basel, Karger, 2011, vol 22, pp 97–114

Mechanisms of Action of Fluoride for Caries Control

Marília Afonso Rabelo Buzalaf[a] · Juliano Pelim Pessan[c] · Heitor Marques Honório[b] · Jacob Martien ten Cate[d]

[a]Department of Biological Sciences and [b]Department of Pediatric Dentistry, Orthodontics and Public Health, Bauru Dental School, University of São Paulo, Bauru, and [c]Department of Pediatric Dentistry and Public Health, Araçatuba Dental School, São Paulo State University, Araçatuba, Brazil; [d]Department of Cariology, Endodontology and Pedodontology, Academic Center for Dentistry Amsterdam, Amsterdam, The Netherlands

Abstract

Fluoride was introduced into dentistry over 70 years ago, and it is now recognized as the main factor responsible for the dramatic decline in caries prevalence that has been observed worldwide. However, excessive fluoride intake during the period of tooth development can cause dental fluorosis. In order that the maximum benefits of fluoride for caries control can be achieved with the minimum risk of side effects, it is necessary to have a profound understanding of the mechanisms by which fluoride promotes caries control. In the 1980s, it was established that fluoride controls caries mainly through its topical effect. Fluoride present in low, sustained concentrations (sub-ppm range) in the oral fluids during an acidic challenge is able to absorb to the surface of the apatite crystals, inhibiting demineralization. When the pH is re-established, traces of fluoride in solution will make it highly supersaturated with respect to fluorhydroxyapatite, which will speed up the process of remineralization. The mineral formed under the nucleating action of the partially dissolved minerals will then preferentially include fluoride and exclude carbonate, rendering the enamel more resistant to future acidic challenges. Topical fluoride can also provide antimicrobial action. Fluoride concentrations as found in dental plaque have biological activity on critical virulence factors of *S. mutans* in vitro, such as acid production and glucan synthesis, but the in vivo implications of this are still not clear. Evidence also supports fluoride's systemic mechanism of caries inhibition in pit and fissure surfaces of permanent first molars when it is incorporated into these teeth pre-eruptively.

Copyright © 2011 S. Karger AG, Basel

The multifactorial disease dental caries is caused by the simultaneous interplay of different factors – dietary sugars, dental biofilm and the host – within the context of the oral environment. The complex and long-lasting interactions of these factors and how they lead to caries was already described half a century ago [1]. With our current understanding, the most obvious way to fight caries is to control the causal agents by removing the dental biofilm and reducing sugar consumption. These approaches form the basis of comprehensive protocols to control the disease, but have been proven insufficient to lead to a desired level of prevention because they strongly rely on patient compliance. Even before a complete understanding of the etiology of dental

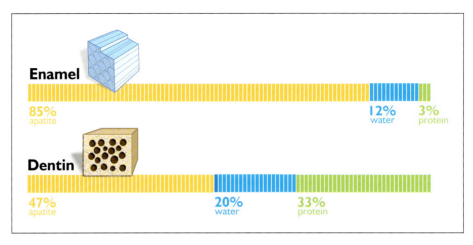

Fig. 1. General composition of dental enamel and dentin.

Biochemistry of Caries Development

Dental caries is the net result of consecutive cycles of de- and remineralization of dental tissues at the interface between the biofilm and

caries was reached, fluoride had emerged as a pivotal adjunct to combat the disease [2]. Fluoride is currently recognized as the main factor responsible for the significant decline in caries prevalence that has been observed worldwide [3]. On the other hand, excessive fluoride intake during the period of tooth development may cause dental fluorosis, the only proven side effect of the use of fluoride of dental relevance [4]. An increase in the prevalence of dental fluorosis has been reported concomitantly with the decrease in caries [5–7]. Although most of this fluorosis is mild or very mild, and has little or no impact on quality of life of affected people [8], a judicious use of fluoride to avoid moderate and severe fluorosis is needed. Thus, in order that the maximum benefit of fluoride for caries control can be achieved with a minimum risk of side effects, it is necessary to have a comprehensive understanding of the mechanisms by which fluoride promotes caries control.

the tooth surface, with demineralization being caused by the production of acids by oral bacteria after sugar consumption [9]. To understand how the acids can attack the dental tissues, it is fundamental to know their biochemical properties.

Composition of Enamel and Dentin

Despite the presence of common constituents, enamel and dentin have different structures that will affect caries progression within these tissues as well as the reactivity of fluoride with them.

Permanent enamel is an acellular tissue composed chiefly of minerals (calcium-deficient carbonated hydroxyapatite, 85% in volume). Hydroxyapatite molecules are arranged in long and thin apatite crystals, which in turn are organized into the resulting enamel prisms (fig. 1). Despite the high mineral content, the space between the crystals is occupied by water (12% by volume) and organic material (3% by volume) [10, 11]. It is in this space filled with the enamel fluid that the de- and remineralization reactions take place. In brief, upon a cariogenic challenge, hydroxyapatite crystals are dissolved from the subsurface, while fluorapatite crystals are deposited at the surface, thus resulting in a subsurface lesion. The dissolution process of enamel is therefore a chemical event.

On the other hand, permanent dentin contains (by volume) 47% apatite, 33% organic components and 20% water (fig. 1). The mineral phase is also hydroxyapatite, similar to enamel, but the crystallites have much smaller dimensions than those found in enamel. As a consequence, the ratio surface area/crystallite volume is larger, which makes the mineral phase more reactive. As a result, dentin surfaces are more susceptible to caries attack than enamel surfaces. The organic matrix is mainly composed of collagen (90%), but there are many non-collagenous components that determine the properties of the matrix and interfere with de- and remineralization reactions. Collagen forms the backbone of dentin and serves as a template for the deposition of apatite crystallites within the collagen helix. This kind of structure promotes a synergism between matrix and apatite: the mineral phase cannot be completely dissolved during an acid attack and the matrix does not undergo enzymatic degradation while its surface is still protected by apatite [11]. Dentin caries is thus a biochemical process characterized initially by the dissolution of the mineral, which in turn exposes the organic matrix to breakdown [12–15] by bacterial-derived enzymes as well as by host-derived enzymes such as matrix metalloproteinases present in dentin and saliva [16, 17]. It is also important to highlight that dentin is a cellular tissue and that upon exogenous challenges the pulpo-dentinal organ responds with mineral deposition [18]. This process, combined with the flow of dentinal fluid from the pulp, reduces the rate of lesion progression in dentin in vivo [19].

Dental Mineral Dynamics

The reason why caries progresses slowly is due to the high supersaturation of saliva with respect to enamel mineral under physiological conditions. This can be easily understood when the concentrations of free ions required to form hydroxyapatite

normally available in saliva are compared with the concentrations that are necessary to reach saturation and form this mineral. The solubility product of enamel (KSP_{enamel}) which is related to the concentrations of Ca^{+2}, PO_4^{-3} and OH^- required for the formation of enamel crystals, has been calculated at 5.5×10^{-55} mol^9/l^9 at 37°C, slightly higher than that required to form hydroxyapatite (KSP_{HA} 7.41×10^{-60} mol^9/l^9). Under physiological conditions (pH 7.0), based on the salivary concentrations of free Ca^{+2}, PO_4^{-3} and OH^- that are available to form enamel crystals, the ion activity product of hydroxyapatite (IAP_{HA}) has been calculated at 6.1×10^{-48} mol^9/l^9 [11]. Therefore, if the IAP_{HA} in saliva under physiological conditions is higher than the concentrations required to form enamel crystals (KSP_{enamel}) this implies that enamel mineral does not dissolve in saliva (fig. 2a). Contrarily, enamel crystals would be expected to grow or new crystals would be expected to form at the biofilm-free tooth surfaces. This does not happen because saliva contains proteins that inhibit hydroxyapatite crystal growth, including statherin and many proline-rich proteins [20].

When a biofilm is covering the enamel surface, it reduces the access of saliva to the tooth. The relevant fluid phase in this case is the biofilm fluid which, under resting conditions, is also supersaturated with respect to enamel (IAP_{HA} 1.4×10^{-47}). This would favor remineralization of previously demineralized enamel or promote the formation of supragingival calculus (fig. 2b).

The characteristics of the plaque fluid microenvironment change considerably upon a sugar challenge. In this case, bacteria produce lactic acid that makes the plaque fluid pH fall (typically between 4.5 and 5.5). The driving force is then shifted to mineral dissolution. But why does this happen if saliva is continuously secreted with relatively stable Ca^{+2} and PO_4^{-3} concentrations, which would apparently maintain IAP_{HAP} unaltered?

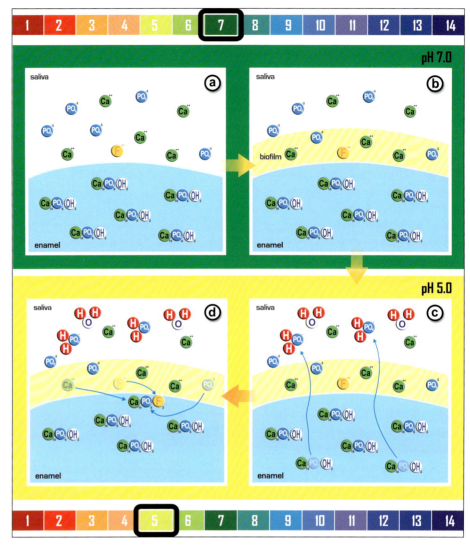

Fig. 2. Dynamics of minerals in saliva and enamel under neutral (**a**, **b**) and acidic conditions (**c**, **d**).

The pH fall has a profound effect on the solubility of hydroxyapatite and other calcium phosphates. In general, the solubility of apatite increases 10 times with a decrease of 1 pH unit. This happens because H^+ combines with PO_4^{-3} and OH^- to form $H_2PO_4^{-3}$ and H_2O (Eq. 1). As a consequence, the concentrations of free PO_4^{-3} and OH^- are reduced, thus decreasing the IAP_{HAP} and turning the solution undersaturated with respect to enamel ($IAP_{HA} < KSP_{enamel}$), promoting enamel dissolution (fig. 2c–d) [11]. The dissolution can be avoided by increasing the concentrations of Ca^{+2} and/or PO_4^{-3} in the fluid. Therefore, the lower the pH, the higher the concentrations

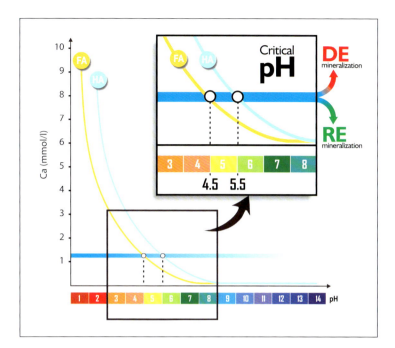

Fig. 3. Solubility of apatite as a function of pH, expressed in terms of calcium concentrations. Blue line indicates salivary calcium concentrations. The critical pH for dissolution of hydroxyapatite (HA) and fluorhydroxyapatite (FA) is 5.5 and 4.5, respectively.

of Ca^{+2} and PO_4^{-3} required to reach saturation in respect to hydroxyapatite. This relationship is shown in figure 3.

$$Ca_5(PO_4)_3OH \rightarrow 5\ Ca^{+2} + 3\ PO_4^{-3} + OH^- \quad (1)$$
$$\downarrow H^+ \qquad \downarrow H^+$$
$$HPO_4^{-2} \quad H_2O$$
$$\downarrow H^+$$
$$H_2PO_4^-$$

When the pH is gradually lowered from 7.0 to 5.0, the value of pH for which the fluid becomes saturated with respect to the mineral in question (IAP = KSP) is the so-called 'critical pH'. At those conditions, equilibrium exists (no mineral dissolution and no mineral precipitation). For hydroxyapatite, the critical pH is around 5.5, while it is approximately 4.5 for fluorhydroxyapatite. When the pH is above the critical level for the formation of a respective mineral phase, precipitation of this phase occurs (remineralization). Contrarily, when the pH is below the critical level, dissolution takes place (demineralization) (fig. 3).

Carious Lesion Formation
The existence of mineral phases with different solubilities in the dental tissues explains the patterns of demineralization found in caries. Under normal conditions (pH around 7.0), the oral fluids are supersaturated with respect to both hydroxyapatite and fluorhydroxyapatite. Thus, there is a tendency towards formation of these two minerals (formation of calculus and remineralization of demineralized areas).

When bacteria metabolize sugars producing lactic acid, pH decreases in saliva and biofilm fluid (4.5<pH<5.5) rendering these fluids undersaturated with respect to hydroxyapatite while still supersaturated with respect to fluorhydroxyapatite. Consequently, hydroxyapatite dissolves from the subsurface and fluorhydroxyapatite forms in the surface layers. Saliva, in turn, has a strong buffering capacity, and this property together with

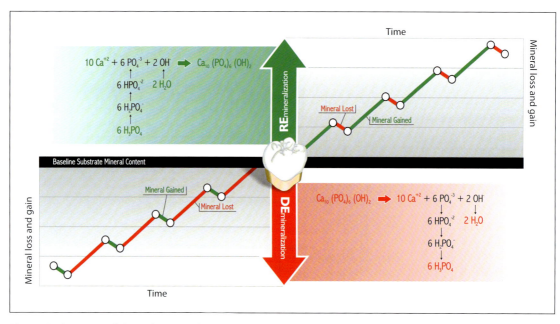

Fig. 4. Cyclic nature of de- and remineralization reactions. Source: Buzalaf et al. [68].

outward diffusion of acids makes the biofilm pH rise within a few minutes. When the pH becomes greater than 5.5, the condition of supersaturation of the oral fluids with respect to hydroxyapatite is restored; the partially demineralized crystals then undergo remineralization. The net result of successive de- and remineralization cycles with the preponderance of the former over the latter leads to caries (fig. 4).

The supersaturation of the oral fluids with respect to fluorhydroxyapatite during cariogenic challenges is responsible for the maintenance of the surface layer of carious lesions (fig. 2d). With time, formation of fluorhydroxyapatite at the expense of hydroxyapatite further increases the concentration of fluorhydroxyapatite in the surface layer. This layer has a protective role, slowing the diffusion of demineralizing agents into the lesion. On the other hand, it also renders remineralization of the lesion body more difficult [11].

Mechanisms by Which Fluoride Controls Caries

Supplementation of public water supplies with controlled levels of fluoride was the first approach involving the use of fluoride for caries control. The encouraging results coming from this measure later prompted the recommendation for the use of fluoride supplements by pregnant women in order to prevent caries in their offspring. Since the first cariostatic benefits of fluoride were observed when this element was ingested from 'systemic' sources, from the 1940s to the 1970s it was originally believed that the cariostatic mechanism of fluoride relied mainly on its uptake in the forming enamel. This would lead to the formation of fluorhydroxyapatite, a mineral phase more resistant to future dissolution. For this purpose, ingestion of fluoride was considered unavoidable and the occurrence of dental fluorosis was regarded as a necessary risk in order to achieve the cariostatic benefits of fluoride.

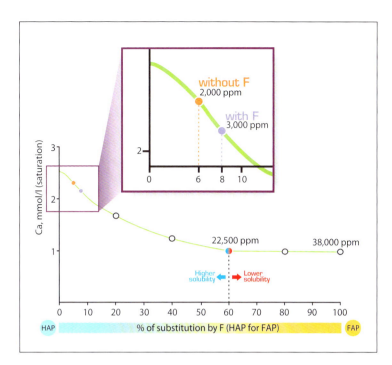

Fig. 5. Calculated solubility of fluorhydroxyapatite at 37°C in 0.1 mol/l acetate buffer at initial pH 5.0 as a function of the degree of replacement of OH⁻ by F⁻.

However, something seemed to be missing. It was observed that fluoride concentrations typically found in enamel were unable to confer significant protection against caries. The highest fluoride concentrations in enamel are found in the surface. They are usually around 2,000 ppm (6% replacement of OH⁻ by F⁻ in hydroxyapatite) in non-fluoridated areas and 3,000 ppm (8% replacement of OH⁻ by F⁻ in hydroxyapatite) in fluoridated areas. However, these concentrations dramatically fall after the outer first 10–20 μm of enamel to around 50 ppm in non-fluoridated areas and hundreds of ppm in fluoridated areas [21]. These levels are far below those able to confer expressive reduction on the solubility of hydroxyapatite (fig. 5).

In the 1980s, the concept that fluoride controls caries lesion development primarily through its topical effect on de- and remineralization processes taking place at the interface between the tooth surface and the oral fluids was established

[22, 23]. Elegant in situ studies conducted in Scandinavia greatly contributed to the consolidation of this concept. In one of the studies, the authors placed human and shark enamel slabs in removable appliances and covered them with orthodontic bands to allow plaque accumulation. Shark enamel was used because it is composed almost of pure fluorapatite (around 30,000 ppm fluoride). Microradiographic analyses revealed that carious lesions formed in both substrates, although they were less severe in shark enamel. The authors compared these data with data from previous studies with human enamel when daily mouthrinsing with 0.2% NaF was used. They observed that the mineral loss in human enamel treated with fluoride rinse was lower than that of shark enamel without any additional treatment. The lesion depths of these substrates were similar (fig. 6) [24]. These studies proved that structurally bound fluoride (shark enamel) was not very effective in inhibiting demineralization, while

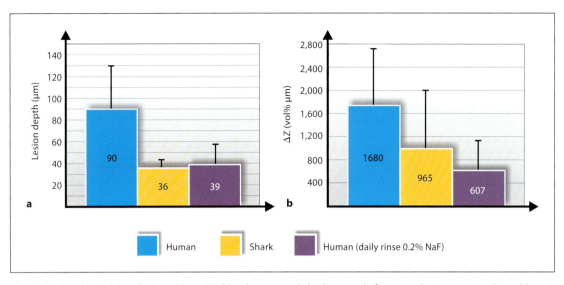

Fig. 6. Lesion depth (**a**) and mineral loss (ΔZ; **b**) in human and shark enamel after 4 weeks in situ as evaluated by microradiography. Groups human and shark refer to human and shark enamel slabs, respectively, which did not receive any additional treatment. Group human (daily rinse 0.2% NaF) refers to human enamel slabs that received daily rinses of 0.2% NaF. Bars indicate SD (n = 6). Original data from Øgaard et al. [24].

fluoride in solution (NaF rinse) led to a high degree of protection. This provided evidence that the primary action of fluoride is topical due to its presence in the fluid phases of the oral environment. It is important to stress out that the concentrations of fluoride found in shark enamel are many times higher than those typically present in human enamel, but even so they were unable to completely inhibit enamel dissolution. On the other hand, fluoride concentrations as little as 1 ppm present in an acid solution can reduce the solubility of carbonated hydroxyapatite to that equivalent to hydroxyapatite. Higher concentrations of fluoride in solution decrease the solubility following a logarithmic pattern [23].

Thus, to interfere in the dynamics of dental caries formation, fluoride must be constantly present in the oral environment at low concentrations. In order that the mechanisms involved in this process can be more easily understood it is helpful initially to consider the different 'pools' of fluoride that can be found in the oral environment.

These pools can be didactically divided into 5 categories [25] (fig. 7):

1   F_O: outer fluoride, present outside enamel (in the biofilm or saliva);
2   F_S: fluoride present in the solid phase, incorporated in the structure of the crystals, also known as fluorhydroxyapatite;
3   F_L: fluoride present at the enamel fluid;
4   F_A: fluoride adsorbed to the crystal surface, also known as loosely-bound;
5   CaF_2: 'CaF$_2$-like' material; globules deposited on enamel and biofilm after application of highly concentrated fluoride products; acts as a pH-controlled fluoride and calcium reservoir.

Fluoride Mechanisms of Action
Inhibition of Demineralization
If fluoride is present in plaque fluid (F_L) when bacteria produce acids, it will penetrate along with the acids at the subsurface, adsorb to the crystal surface (F_A) and protect crystals from dissolution [26]. When the entire crystal surface is covered

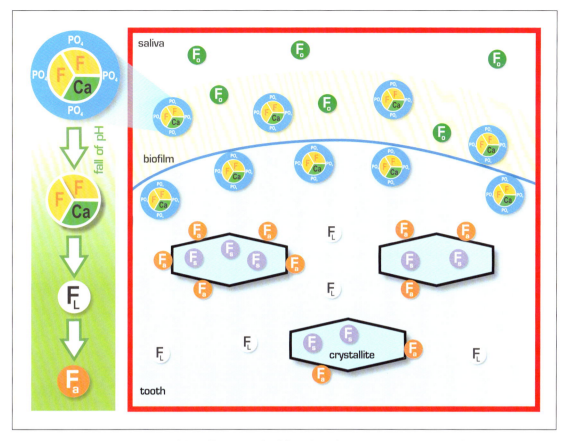

Fig. 7. Schematic representation of the different 'pools' of fluoride in the oral environment. Modified from Arends and Christoffersen [25].

by F_A (100% coverage), it will not dissolve upon a pH fall caused by bacterial-derived acids, since this type of coating makes the characteristics of the crystal similar to those of fluorapatite. On the other hand, when the coating of F_A is partial, the uncoated parts of the crystal will undergo dissolution (fig. 8) [25].

While F_A is the 'pool' of fluoride that effectively protects the crystals from dissolution, the role of fluoride present in solution (F_L) is equally important, since the higher the concentration of F_L, the higher the probability that it adsorbs (F_A) and protects the crystals. However, very low fluoride concentrations (sub-ppm range) in solution are

already able to substantially inhibit acid dissolution of tooth minerals [23, 27].

Calcium fluoride (CaF_2) is an important source of fluoride to the oral fluids (F_L). It is known as pH-controlled fluoride and calcium reservoir. This compound forms when the fluoride concentrations in the solution bathing enamel are higher than 100 ppm. The formation of CaF_2 is a two-stage reaction. Initially, a slight dissolution of the enamel surface must occur to release Ca^{+2} that in a second stage will react with fluoride that is applied, thereby forming CaF_2 globules. These globules precipitate not only on sound enamel surfaces but also and more importantly on biofilm,

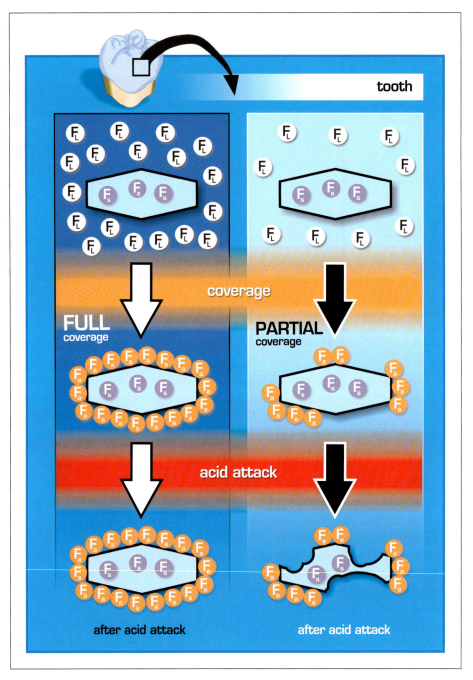

Fig. 8. Events taking place at the subsurface of enamel upon a cariogenic acidic challenge. Fluoride (F_L) penetrates at the subsurface along with the acids, adsorbs to the surface of the crystal and protects it from dissolution (left chart). When coverage is partial, uncovered portions of the crystal will dissolve (right chart). Modified from Arends and Christoffersen [25].

Buzalaf · Pessan · Honório · ten Cate

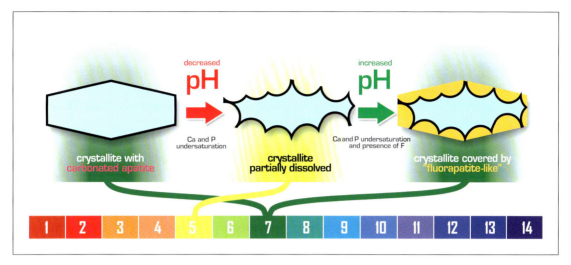

Fig. 9. Schematic representation of remineralization occurring in the presence of fluoride. Fluoride speeds up the process of remineralization and leads to the precipitation of a coat poor in carbonate and rich in fluoride on the partially demineralized original crystallite. This renders the tooth structure more resistant to subsequent acidic challenges. Modified from Featherstone [26].

pellicle and enamel porosities. The dissolution rate of CaF_2 globules is limited by the adsorption of HPO_4^{-2} that is lost under acidic pH, thus allowing CaF_2 to dissolve and fluoride and calcium to be released. This fluoride will add to the 'pool' of F_L [11, 28].

Enhancement of Remineralization

After an acidic challenge, salivary flow buffers the acids produced by the bacteria. When the pH is higher than 5.5, remineralization will naturally occur (fig. 3) since saliva is supersaturated with respect to the dental mineral. Traces of fluoride in solution during dissolution of hydroxyapatite will make the solution highly supersaturated with respect to fluorhydroxyapatite. This will speed up the process of remineralization. Fluoride will adsorb to the surface of the partially demineralized crystals and attract calcium ions. Since carbonate-free or low-carbonate apatite is less soluble, these phases will tend to form preferentially instead of the original mineral, under the nucleating action of the partially dissolved minerals. This new

coating will be less soluble due to the exclusion of carbonate and incorporation of fluoride, rendering the enamel more resistant to future acidic challenges (fig. 9). After repeated cycles of dissolution and reprecipitation, enamel crystals may be completely different from their original state [11, 26].

Role of 'Systemic' Fluoride

As mentioned above, the main mechanisms of action of fluoride rely on its topical use since low, sustained levels of fluoride in the oral fluids can significantly control caries progression and reversal. However, this concept does not invalidate the use of 'systemic' methods such as fluoridated water. More than 60 years of intensive research attest to the safety and effectiveness of this measure to control caries [4]. In this case, however, it should be emphasized that despite being classified as a 'systemic' method of fluoride delivery (as it involves ingestion of fluoride), the mechanism of action of fluoridated water to control caries is mainly through its topical contact with the teeth while in the oral cavity or when redistributed to the oral environment by means of saliva. Since fluoridated water is consumed many times a day, the high frequency of contact of fluoride present in the water with the tooth structure or intraoral fluoride reservoirs helps to explain why water fluoridation is so effective in controlling caries, despite having fluoride concentrations much lower than fluoride toothpastes, for example [29]. This general concept can be applied to all methods of fluoride use traditionally classified as 'systemic'. In the light of the current knowledge regarding the mechanisms by which fluoride control caries, this system of classification is in fact misleading.

One point that deserves attention regarding the mechanism by which fluoridated water leads to caries control is that even recent studies have shown a beneficial pre-eruptive effect of water fluoride on caries control. Well-designed cohort studies have reported that pre-eruption exposure to fluoride is important for caries prevention, especially in pit and fissure surfaces of permanent first molars. This could be due to the difficult access of topical fluoride to these areas. The anticaries protection may occur due to pre-eruption fluoride uptake in the crystalline structure (F_S) of the developing enamel, its adsorption on the crystal surface (F_A) or its presence in the enamel fluid (F_L). Upon post-eruption acidic challenge, F_S would be released to the fluid phase (F_L), thus inhibiting demineralization and enhancing remineralization [30, 31].

Effects in Oral Bacteria

Although the main action of fluoride on the dynamics of dental caries is on de- and remineralization processes that occur on dental hard tissues, it has also been proposed that the fluoride ion can affect the physiology of microbial cells, including cariogenic streptococci, which can thus indirectly affect demineralization [32, 33]. The inhibitory effect of fluoride in pure cultures of oral streptococci was described over 70 years ago, and

since then many reports have been published on direct and indirect effects of fluoride on the energy and biosynthesis of streptococci [34]. Bacterial metabolism can be affected by fluoride through several complex mechanisms that are beyond the scope of the present chapter and therefore will be presented only briefly.

Fluoride exerts its effects on oral bacteria by a direct inhibition of cellular enzymes (directly or in combination with metals) or enhancing proton permeability of cell membranes in the form of hydrogen fluoride (HF) [33, 35]. The biological effects and mechanisms of action of fluoride on oral bacteria are summarized in table 1.

According to the reaction $H^+ + F^- \rightleftharpoons HF$, HF is formed more easily under acidic conditions ($pK_a = 3.15$) and enters the cell due to a higher permeability of HF to bacterial cell membranes. HF then dissociates in H^+ and F^- in the cytoplasm, which is more alkaline than the exterior environment [34]. This intracellular F^- inhibits glycolytic enzymes, resulting in a decrease in acid production from glycolysis. F^- in the cytoplasm also lowers cytoplasmatic pH (which decreases the entire glycolytic activity), affecting both the acid production and acid-tolerance of S. mutans [33]. Cell membrane-associated H^+-ATPases are also inhibited by F^- because excreted protons are brought back into the cell, therefore decreasing excretion of H^+ from the cell (fig. 10) [35, 36].

It is known that fluoride concentrations in plaque can be increased for several hours after exposure to a fluoridated dentifrice [37–40]. Lynch et al. [41] concluded that low levels of plaque and salivary fluoride resulting from the use of 1,500 ppm fluoride toothpastes are insufficient to have a significant antimicrobial effect on plaque bacteria. A recent review, however, concluded that fluoride concentrations as found in dental plaque have biological activity on critical virulence factors of S. mutans in vitro, such as acid production and glucan synthesis, but the in vivo implications are still not clear [33].

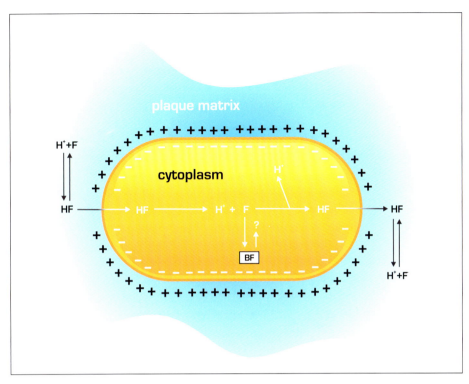

Fig. 10. Fluoride accumulation, distribution and efflux from bacterial cells. BF=Bound fluoride. Modified from Hamilton and Bowden [69].

Table 1. Biological effects and mechanisms of action of fluoride on oral bacteria

Biological activity	Examples	Mechanism
Enzyme inhibition (at sub-millimolar levels of fluoride)	enolase, urease, P-ATPase, phosphatases, heme catalase, heme peroxidase	direct binding of F⁻ or HF
	F-ATPase, nitrogenase, RecA, CheY	binding of metal-F complex
Dissipation of proton gradient/ motive force (at micromolar levels of fluoride)	acidification of cytoplasm (inhibition of glycolysis, PTS system, and IPS formation)	action as transmembrane proton carrier
	inhibition of macromolecular synthesis and export	

PTS system = Phosphotransferase sugar transport system; IPS formation = intracellular polysaccharide formation.
Source: Koo [33].

As most of the evidence of antimicrobial effects of fluoride on oral bacteria comes from in vitro studies, caution must be taken when interpreting these results. Clinical studies addressing the subject, however, seem to indicate that fluoride does have an antimicrobial effect, and that this effect is dependent on factors such as the fluoride concentration applied and associated antibacterial components. With regard to fluoride concentration, studies with different research protocols have shown significantly lower plaque scores in subjects using a 5,000 ppm fluoride toothpaste, in comparison with formulations containing 500, 1,100 and 1,500 ppm fluoride [42, 43]. Concerning other components with inhibitory effects on plaque growth, it was also demonstrated that the combination of high levels of fluoride (5,000 ppm) and sodium lauryl sulphate reduces de novo plaque formation in subjects using slurries of dentifrices with different fluoride concentrations [43]. Also, the association of fluoride with other ions in formulations containing stannous fluoride or amine fluoride has been shown to be effective in promoting lower plaque formation and acid production, either alone or in combination [44–46]. The use of a stabilized stannous fluoride/sodium hexametaphosphate dentifrice [47, 48] as well as a stannous-containing sodium fluoride dentifrice [49] have also proven to be effective in reducing plaque formation.

Fluoride-releasing materials have also been shown to provide antimicrobial effects. Results from in vitro and in situ studies indicate that fluoride released from glass ionomer cements has an inhibitory effect on the pH fall and the acid production rate of S. mutans and S. sanguinis [36]. Reduced S. mutans growth and lower pH fall on plaque formed on glass ionomer cements has also been shown to occur when compared with composite resin [50–52].

Fluoride in Intraoral Reservoirs
Besides interfering in de- and remineralization processes, along with effects in oral bacteria, fluoride retained in intraoral reservoirs plays an important role on the mechanism of action of the ion. It is known that plaque and salivary fluoride levels decrease rapidly after the application of a fluoride vehicle, following a bi-phasic exponential pattern [53]. These levels, however, are significantly elevated for many hours after the exposure to the fluoridated agent when compared to baseline levels, indicating that fluoride is bound to intraoral reservoirs and subsequently released to saliva over time [29, 37–40].

Fluoride can be deposited on dental hard tissues as CaF_2 (as discussed above), bound to the oral mucosa and retained by dental plaque components. Oral mucosa has been shown to be an important fluoride reservoir, mainly due to its large surface area, releasing fluoride to saliva over time [54]. Although all fluoride reservoirs contribute to the maintenance of the ion in the oral cavity, fluoride retained in dental plaque is likely more relevant from a clinical perspective [for details, see Vogel, this vol., pp. 146–157], as it is the site where de- and remineralization processes take place. Considering that most subjects do not completely remove dental plaque after toothbrushing, the amount of fluoride retained in plaque can help determine the fate of the enamel underneath it [37–39].

Fluoride has a strong affinity to both organic and inorganic components of plaque, and can be found as ionic, ionizable and strongly bound forms. Although the amount of fluoride in the ionizable fraction is considerably larger than in the ionic pool, it adds to the amount of ionic fluoride in plaque fluid, which is responsible for the cariostatic action of fluoride [29]. The clinical relevance of fluoride retained in plaque is that it can be released under acidic conditions during cariogenic challenges. In other words, fluoride is released when it is most needed to reduce demineralization, to enhance remineralization of early lesions, or both. Clinical studies support the concept that the amount of fluoride in oral reservoirs is of paramount importance in its cariostatic effectiveness, as caries incidence and activity have been shown to be inversely related to fluoride

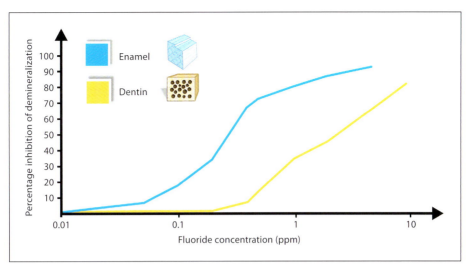

Fig. 11. Inhibition of demineralization of enamel and dentin at different concentrations of fluoride in solution. Data are expressed as percentage of demineralization at 0 ppm fluoride. Modified from ten Cate et al. [11].

concentrations in saliva and/or dental plaque [55–57].

Dentin De- and Remineralization and the Protective Effect of Fluoride

The essence of de- and remineralization processes, as well as the interactions with fluoride that were described above for enamel, also apply to dentin. The main differences of both substrates are:

(1) Dentin is more susceptible to caries attack than enamel, with a critical pH more than 1 pH unit higher than that for enamel [58].

(2) Dentin demineralizes faster and remineralizes slower than enamel under the same experimental conditions [59, 60].

(3) More concentrated fluoride is needed to inhibit demineralization [61, 62] (fig. 11) and to enhance remineralization [63] of dentin when compared with enamel. In fact, clinical trials show a beneficial effect of 5,000 ppm fluoride over 1,100

ppm fluoride dentifrices to arrest root carious lesions [42, 64].

(4) Dentin seems to benefit from a higher daily frequency of exposure to fluoride [65] and also from the combination of methods of fluoride use [66] which is not necessarily the case for enamel.

(5) Dentin contact area with cariogenic acids is larger than that of enamel. For this reason, dentin is apparently much more permeable to acids, with demineralization taking place at a relatively large depth, while mineral deposition is restricted to the outer layers. If the crystallites surrounding the diffusion channels (tubules) are coated with a fluoride-rich mineral, the acids will bypass these relatively resistant minerals, while mineral and fluoride ions will readily be deposited. Thus, the lesion front in dentin moves deeper, while the surface layer becomes broader. In enamel, on the other hand, diffusion is much slower and allows acids to 'sidestep' into smaller intraprismatic porosities and dissolve crystallites that are still unaffected by either acid or fluoride. Thus, mineral uptake and loss occur at similar depths for enamel

lesions, while for dentin lesions mineral uptake is predominant at the surface and mineral loss at the lesion front [60].

It has also been recently shown that very deep lesions extending through enamel into dentin can be remineralized. Although this process is slow, it indicates that remineralization might be used to treat deep lesions [67].

Conclusion

Knowledge of the mechanisms by which fluoride promotes caries control is essential for the achievement of the maximum benefits of this element with minimum risk of side effects. The main action of fluoride for caries control occurs through its topical effect. Fluoride present in low, sustained concentrations (sub-ppm range) in the oral fluids during an acidic challenge is able to absorb to the surface of the apatite crystals, inhibiting demineralization. When the pH is reestablished, traces of fluoride in solution will make it highly supersaturated with respect to fluorhydroxyapatite, which will speed up the process of remineralization. The mineral formed under the nucleating action of the partially dissolved minerals will then preferentially include fluoride and exclude carbonate, rendering the enamel more resistant to future acidic challenges. Topical fluoride can also present antimicrobial action. Fluoride concentrations as found in dental plaque have biological activity on critical virulence factors of S. mutans in vitro, such as acid production and glucan synthesis, but the in vivo implications of this are still not clear.

Evidence from cohort studies also supports fluoride's systemic mechanism of caries inhibition in pit and fissure surfaces of permanent first molars when it is incorporated into these teeth pre-eruptively. In this case, upon post-eruption acidic challenge, F_S would be released to the fluid phase (F_L), thus inhibiting demineralization and enhancing remineralization. Additionally, ingested fluoride can exert a topical mechanism of action when it recirculates in the oral environment through saliva.

The essence of de- and remineralization processes, as well as the interactions with fluoride that were described for enamel, also apply to dentin. The main differences are that dentin is more susceptible to caries attack than enamel, with a critical pH more than 1 pH unit higher. Consequently, dentin demineralizes faster and remineralizes slower, requiring higher fluoride concentrations and frequencies of application when compared with enamel.

References

1 Keyes PH: The infectious and transmissible nature of experimental dental caries: findings and implications. Arch Oral Biol 1960;1:304–320.
2 Dean HT, Arnold FA, Elvolve E: Additional studies of the relation of fluoride domestic waters to dental caries experience in 4,425 white children aged 12–14 years in 13 cities in 4 states. Public Health Rep 1942;57:1155–1179.
3 Bratthall D, Hansel-Petersson G, Sundberg H: Reasons for the caries decline: what do the experts believe? Eur J Oral Sci 1996;104:416–422; discussion 423–425, 430–432.
4 McDonagh MS, Whiting PF, Wilson PM, Sutton AJ, Chestnutt I, Cooper J, Misso K, Bradley M, Treasure E, Kleijnen J: Systematic review of water fluoridation. BMJ 2000;321:855–859.
5 Clark DC: Trends in prevalence of dental fluorosis in North America. Community Dent Oral Epidemiol 1994;22:148–152.
6 Khan A, Moola MH, Cleaton-Jones P: Global trends in dental fluorosis from 1980 to 2000: a systematic review. SADJ 2005;60:418–421.
7 Whelton HP, Ketley CE, McSweeney F, O'Mullane DM: A review of fluorosis in the European Union: prevalence, risk factors and aesthetic issues. Community Dent Oral Epidemiol 2004;32(suppl 1): 9–18.
8 Chankanka O, Levy SM, Warren JJ, Chalmers JM: A literature review of aesthetic perceptions of dental fluorosis and relationships with psychosocial aspects/oral health-related quality of life. Community Dent Oral Epidemiol 2010;38: 97–109.

9 Fejerskov O, Kidd EA, Nyvad B, Baelum V: Defining the disease: an introduction; in Fejerskov O, Kidd E (eds): Dental Caries The Disease and its Clinical Management, ed 2. Oxford, Blackwell Munksgaard, 2008, pp 3–6.

10 Featherstone JD, Lussi A: Understanding the chemistry of dental erosion. Monogr Oral Sci 2006;20:66–76.

11 ten Cate JM, Larsen MJ, Pearce EIF, Fejerskov O: Chemical interactions between the tooth and oral fluids; in Fejerskov O, Kidd E (eds): Dental Caries the Disease and Its Clinical Management. Oxford, Blackwell Munksgaard, 2008, pp 209–231.

12 Klont B, ten Cate JM: Remineralization of bovine incisor root lesions in vitro: the role of the collagenous matrix. Caries Res 1991;25:39–45.

13 Kleter GA, Damen JJ, Everts V, Niehof J, ten Cate JM: The influence of the organic matrix on demineralization of bovine root dentin in vitro. J Dent Res 1994;73: 1523–1529.

14 Nyvad B, Fejerskov O: An ultrastructural study of bacterial invasion and tissue breakdown in human experimental root-surface caries. J Dent Res 1990;69: 1118–1125.

15 Schupbach P, Guggenheim B, Lutz F: Human root caries: histopathology of initial lesions in cementum and dentin. J Oral Pathol Med 1989;18:146–156.

16 Tjäderhane L, Larjava H, Sorsa T, Uitto VJ, Larmas M, Salo T: The activation and function of host matrix metalloproteinases in dentin matrix breakdown in caries lesions. J Dent Res 1998;77: 1622–1629.

17 Chaussain-Miller C, Fioretti F, Goldberg M, Menashi S: The role of matrix metalloproteinases (MMPs) in human caries. J Dent Res 2006;85:22–32.

18 Frank RM, Voegel JC: Ultrastructure of the human odontoblast process and its mineralisation during dental caries. Caries Res 1980;14:367–380.

19 Shellis RP: Effects of a supersaturated pulpal fluid on the formation of caries-like lesions on the roots of human teeth. Caries Res 1994;28:14–20.

20 Moreno EC, Varughese K, Hay DI: Effect of human salivary proteins on the precipitation kinetics of calcium phosphate. Calcif Tissue Int 1979;28:7–16.

21 Weatherell JA, Deutsch D, Robinson C, Hallsworth AS: Assimilation of fluoride by enamel throughout the life of the tooth. Caries Res 1977;11(suppl 1): 85–115.

22 ten Cate JM: In vitro studies on the effects of fluoride on de- and remineralization. J Dent Res 1990;69 Spec No:614–9; discussion 34–6.

23 Featherstone JD, Glena R, Shariati M, Shields CP: Dependence of in vitro demineralization of apatite and remineralization of dental enamel on fluoride concentration. J Dent Res 1990;69 Spec No:620–625, discussion 634–636.

24 Øgaard B, Rølla G, Ruben J, Dijkman T, Arends J: Microradiographic study of demineralization of shark enamel in a human caries model. Scand J Dent Res 1988;96:209–211.

25 Arends J, Christoffersen J: Nature and role of loosely bound fluoride in dental caries. J Dent Res 1990;69 Spec No: 601–605, discussion 634–636.

26 Featherstone JD: Prevention and reversal of dental caries: role of low level fluoride. Community Dent Oral Epidemiol 1999;27:31–40.

27 ten Cate JM, Featherstone JD: Mechanistic aspects of the interactions between fluoride and dental enamel. Crit Rev Oral Biol Med 1991;2:283–296.

28 Rølla G: On the role of calcium fluoride in the cariostatic mechanism of fluoride. Acta Odontol Scand 1988;46:341–345.

29 Whitford GM, Wasdin JL, Schafer TE, Adair SM: Plaque fluoride concentrations are dependent on plaque calcium concentrations. Caries Res 2002;36: 256–265.

30 Singh KA, Spencer AJ: Relative effects of pre- and post-eruption water fluoride on caries experience by surface type of permanent first molars. Community Dent Oral Epidemiol 2004;32:435–446.

31 Singh KA, Spencer AJ, Brennan DS: Effects of water fluoride exposure at crown completion and maturation on caries of permanent first molars. Caries Res 2007;41:34–42.

32 Marquis RE: Antimicrobial actions of fluoride for oral bacteria. Can J Microbiol 1995;41:955–964.

33 Koo H: Strategies to enhance the biological effects of fluoride on dental biofilms. Adv Dent Res 2008;20:17–21.

34 ten Cate JM, van Loveren C: Fluoride mechanisms. Dent Clin North Am 1999;43:713–42, vii.

35 Marquis RE, Clock SA, Mota-Meira M: Fluoride and organic weak acids as modulators of microbial physiology. FEMS Microbiol Rev 2003;26:493–510.

36 Nakajo K, Imazato S, Takahashi Y, Kiba W, Ebisu S, Takahashi N: Fluoride released from glass-ionomer cement is responsible to inhibit the acid production of caries-related oral streptococci. Dent Mater 2009;25:703–708.

37 Pessan JP, Alves KM, Ramires I, et al.: Effects of regular and low-fluoride dentifrices on plaque fluoride. J Dent Res 2010;89:1106–1110.

38 Pessan JP, Sicca CM, de Souza TS, da Silva SM, Whitford GM, Buzalaf MA: Fluoride concentrations in dental plaque and saliva after the use of a fluoride dentifrice preceded by a calcium lactate rinse. Eur J Oral Sci 2006;114:489–493.

39 Pessan JP, Silva SM, Lauris JR, Sampaio FC, Whitford GM, Buzalaf MA: Fluoride uptake by plaque from water and from dentifrice. J Dent Res 2008;87:461–465.

40 Whitford GM, Buzalaf MA, Bijella MF, Waller JL: Plaque fluoride concentrations in a community without water fluoridation: effects of calcium and use of a fluoride or placebo dentifrice. Caries Res 2005;39:100–107.

41 Lynch RJ, Navada R, Walia R: Low-levels of fluoride in plaque and saliva and their effects on the demineralisation and remineralisation of enamel; role of fluoride toothpastes. Int Dent J 2004;54: 304–309.

42 Baysan A, Lynch E, Ellwood R, Davies R, Petersson L, Borsboom P: Reversal of primary root caries using dentifrices containing 5,000 and 1,100 ppm fluoride. Caries Res 2001;35:41–46.

43 Nordstrom A, Mystikos C, Ramberg P, Birkhed D: Effect on de novo plaque formation of rinsing with toothpaste slurries and water solutions with a high fluoride concentration (5,000 ppm). Eur J Oral Sci 2009;117:563–567.

44 Madlena M, Dombi C, Gintner Z, Banoczy J: Effect of amine fluoride/stannous fluoride toothpaste and mouthrinse on dental plaque accumulation and gingival health. Oral Dis 2004;10:294–297.

45 Gerardu VA, van Loveren C, Heijnsbroek M, Buijs MJ, van der Weijden GA, ten Cate JM: Effects of various rinsing protocols after the use of amine fluoride/stannous fluoride toothpaste on the acid production of dental plaque and tongue flora. Caries Res 2006;40:245–250.

46 Paraskevas S, van der Weijden GA: A review of the effects of stannous fluoride on gingivitis. J Clin Periodontol 2006;33: 1–13.

47 Bellamy PG, Jhaj R, Mussett AJ, Barker ML, Klukowska M, White DJ: Comparison of a stabilized stannous fluoride/ sodium hexametaphosphate dentifrice and a zinc citrate dentifrice on plaque formation measured by digital plaque imaging (DPIA) with white light illumination. J Clin Dent 2008;19:48–54.

48 Bellamy PG, Khera N, Day TN, Barker ML, Mussett AJ: A randomized clinical trial to compare plaque inhibition of a sodium fluoride/potassium nitrate dentifrice versus a stabilized stannous fluoride/sodium hexametaphosphate dentifrice. J Contemp Dent Pract 2009;10:1–9.

49 He T, Sun L, Li S, Ji N: The anti-plaque efficacy of a novel stannous-containing sodium fluoride dentifrice: a randomized and controlled clinical trial. Am J Dent 2010;23:11–16.

50 Benelli EM, Serra MC, Rodrigues AL Jr, Cury JA: In situ anticariogenic potential of glass ionomer cement. Caries Res 1993;27:280–284.

51 Seppa L, Korhonen A, Nuutinen A: Inhibitory effect on S. mutans by fluoride-treated conventional and resin-reinforced glass ionomer cements. Eur J Oral Sci 1995;103:182–185.

52 Seppa L, Torppa-Saarinen E, Luoma H: Effect of different glass ionomers on the acid production and electrolyte metabolism of Streptococcus mutans Ingbritt. Caries Res 1992;26:434–438.

53 ten Cate JM: Current concepts on the theories of the mechanism of action of fluoride. Acta Odontol Scand 1999;57: 325–329.

54 Zero DT, Raubertas RF, Pedersen AM, Fu J, Hayes AL, Featherstone JD: Studies of fluoride retention by oral soft tissues after the application of home-use topical fluorides. J Dent Res 1992;71: 1546–1552.

55 Gaugler RW, Bruton WF: Fluoride concentration in dental plaque of naval recruits with and without caries. Arch Oral Biol 1982;27:269–272.

56 Schamschula RG, Sugar E, Un PS, Toth K, Barmes DE, Adkins BL: Physiological indicators of fluoride exposure and utilization: an epidemiological study. Community Dent Oral Epidemiol 1985;13: 104–107.

57 Nobre dos Santos M, Melo dos Santos L, Francisco SB, Cury JA: Relationship among dental plaque composition, daily sugar exposure and caries in the primary dentition. Caries Res 2002;36:347–352.

58 Hoppenbrouwers PM, Driessens FC, Borggreven JM: The demineralization of human dental roots in the presence of fluoride. J Dent Res 1987;66:1370–1374.

59 Arends J, Christoffersen J, Buskes JA, Ruben J: Effects of fluoride and methanehydroxydiphosphate on enamel and on dentine demineralization. Caries Res 1992;26:409–417.

60 ten Cate JM, Buijs MJ, Damen JJ: pH-cycling of enamel and dentin lesions in the presence of low concentrations of fluoride. Eur J Oral Sci 1995;103: 362–367.

61 ten Cate JM, Duijsters PP: Influence of fluoride in solution on tooth demineralization. I. Chemical data. Caries Res 1983;17:193–199.

62 ten Cate JM, Damen JJ, Buijs MJ: Inhibition of dentin demineralization by fluoride in vitro. Caries Res 1998;32: 141–147.

63 Herkstroter FM, Witjes M, Arends J: Demineralization of human dentine compared with enamel in a pH-cycling apparatus with a constant composition during de- and remineralization periods. Caries Res 1991;25:317–322.

64 Lynch E, Baysan A, Ellwood R, Davies R, Petersson L, Borsboom P: Effectiveness of two fluoride dentifrices to arrest root carious lesions. Am J Dent 2000;13: 218–220.

65 Laheij AM, van Strijp AJ, van Loveren C: In situ remineralisation of enamel and dentin after the use of an amine fluoride mouthrinse in addition to twice daily brushings with amine fluoride toothpaste. Caries Res 2010;44:260–266.

66 Vale GC, Tabchoury CP, Del Bel Cury AA, Tenuta LM, Ten Cate JM, Cury JA: APF and dentifrice effect on root dentine demineralization and biofilm. J Dent Res 2011;90:77–81.

67 ten Cate JM: Remineralization of deep enamel dentine caries lesions. Aust Dent J 2008;53:281–285.

68 Buzalaf MA, Hannas AR, Magalhaes AC, Rios D, Honorio HM, Delbem AC: pH-cycling models for in vitro evaluation of the efficacy of fluoridated dentifrices for caries control: strengths and limitations. J Appl Oral Sci 2010;18:316–334.

69 Hamilton IR, Bowden GHW: Fluoride effects on oral bacteria; in Fejerskov O, Ekstrand J, Burt BA (eds): Fluoride in Dentistry, ed 2. Copenhagen, Muksgaard, 1996, pp 230–251.

Marília Afonso Rabelo Buzalaf
Department of Biological Sciences
Bauru Dental School, University of São Paulo
Al. Octávio Pinheiro Brisolla, 9–75
Bauru-SP, 17012–901 (Brazil)
Tel. +55 14 3235 8346, E-Mail mbuzalaf@fob.usp.br

Buzalaf MAR (ed): Fluoride and the Oral Environment.
Monogr Oral Sci. Basel, Karger, 2011, vol 22, pp 115–132

Topical Use of Fluorides for Caries Control

Juliano Pelim Pessan[a] · Kyriacos Jack Toumba[c] ·
Marília Afonso Rabelo Buzalaf[b]

[a]Department of Pediatric Dentistry and Public Health, Araçatuba Dental School, São Paulo State University, Araçatuba, and
[b]Department of Biological Sciences, Bauru Dental School, University of São Paulo, Bauru, Brazil; [c]Division of Child Dental Health,
Leeds Dental Institute, University of Leeds, Leeds, UK

Abstract

Since the early findings on the protective effects of fluoride present in drinking water upon caries incidence
and prevalence, intensive research has been conducted
in order to determine the benefits, safety, as well as the
cost-effectiveness of other modalities of fluoride delivery. The present chapter reviews the various forms of topical fluoride use – professionally and self-applied – with
special emphasis on clinical efficacy and possible side
effects. The most widely used forms of fluoride delivery
have been subject of several systematic reviews, providing strong evidence supporting the use of dentifrices,
gels, varnishes and mouth rinses for the control of caries progression. Dentifrices with fluoride concentrations
of 1,000 ppm and above have been shown to be clinically effective in caries prevention when compared to a
placebo treatment, but the evidence regarding formulations with 450–550 ppm is still subject of debate. Therefore, the recommendation for low-fluoride dentifrice use
must take into account both risks and benefits. The evidence for the combined use of two modalities of fluoride application in comparison to a single modality is still
inconsistent, implying that more studies with adequate
methodology are needed to determine the real benefits
of each method. Considering the currently available evidence and risk-benefit aspects, it seems justifiable to recommend the use of fluoridated dentifrices to individuals
of all ages, and additional fluoride therapy should also be
targeted towards individuals at high caries risk.

Since the early findings of the protective effects
of fluoride present in drinking water upon caries incidence and prevalence over 70 years ago,
intensive research has been conducted in order to
determine the benefits, safety, as well as the cost-
effectiveness of modalities of fluoride delivery
other than water. These include both topical and
systemic methods of fluoride delivery, which differ widely regarding fluoride concentration and
mode of application. The current scientific consensus regards a constant supply of low levels of
fluoride, especially at the biofilm/saliva/dental
interface (topical effect), as being the most beneficial in preventing dental caries [1]. Therefore,
the classification of the methods of fluoride delivery into *topical* and *systemic* has been recently
questioned. Ellwood et al. [2] proposed that the
methods should be classified according to their
mode of application, as follows:

– *Community Methods:* introduced on a
population basis (water, milk and salt
fluoridation);

– *Self-Applied Methods:* used at home (toothpastes, mouth rinses, tablets, drops, lozenges and chewing gums);
– *Professional Methods:* delivered by healthcare professionals (solutions, gels, foams, varnishes, slow-release fluoride devices and fluoride-releasing dental materials).

However, for didactical reasons, the present chapter will use the term *topical fluorides* to describe the methods that provide fluoride to exposed surfaces of the dentition at elevated concentrations for a local protective effect, and which are therefore not intended for ingestion [3]. These include both self-applied and professional methods of fluoride delivery. Community methods are be discussed in the paper by Sampaio and Levy [this vol., pp. 133–145].

It is important to highlight that most of the scientific knowledge on the effectiveness of topical fluoride methods was obtained when caries incidence and prevalence were still very high, so even modest interventions led to significant reductions in caries levels [4]. As a substantial decrease in caries prevalence has been observed over the last decades, authorities have questioned the real benefit of topical fluoride applications. A large number of clinical trials have been conducted in order to assess the clinical effectiveness of the various modalities of topical fluoride administration, using different protocols that are not always appropriate and do not allow direct comparisons between results. In order to avoid misinterpretation of the information from those trials due to methodological issues, meta-analytical approaches have been used in systematic reviews aiming to provide real estimates of the effectiveness and safety of the various forms of topical fluoride administration. In the present chapter, the main conclusions of the most recent systematic reviews conducted by the Cochrane Central Register of Controlled Trials (the most comprehensive source of reports of trials available) [5] is presented whenever evidence is available, so professionals and patients can base their choice of a certain fluoride vehicle – or combination of modalities – on scientific knowledge rather than just personal preference. It is worth mentioning that systematic reviews usually adopt the prevented fraction as the primary estimate of effect of the treatment, which refers to the caries increment in the treatment group expressed as a percentage of the control group [6]. This estimate has also been adopted in the present review.

Fluoride Compounds

There is a diversity of fluoride compounds used in fluoride agents available to the public and to healthcare professionals. These differ greatly according to their chemical structures, which ultimately have implications on the mode of action for each compound. According to Axelsson [7], the three main categories are:

1 inorganic compounds: readily soluble salts that provide free fluoride;
2 monofluorophosphate-containing compounds: fluoride is covalently bound to PO_3^{2-} ions and requires hydrolysis to release fluoride ions;
3 organic fluorides: fluoride bound to organic compounds.

The main fluoride compounds, along with their main characteristics and vehicles in which they are frequently used, are briefly summarized in table 1. It is worth mentioning that some compounds can be used in combination in different vehicles, such as toothpastes (NaF/MFP, SnF_2/AmF), prophylaxis pastes (NaF/MFP, NaF/AmF) and mouth rinses (SnF_2/AmF) [7].

Solutions

Fluoridated solutions were the first vehicles of professionally applied fluorides aiming at a reduction in caries levels [8]. These included neutral NaF solutions, as well as SnF_2, acidulated

Table 1. Main fluoride compounds used in topical formulations

Compounds	Characteristics	Vehicles in which the compound is used
Inorganic compounds		
Sodium fluoride (NaF)	the most commonly used fluoride compound (both self-application and professional use); when in solution, NaF salt readily releases fluoride into saliva, dental plaque, pellicle and enamel crystallites	dentifrices mouth rinses chewing-gums solutions gels varnishes prophylaxis pastes slow-release devices
Stannous fluoride (SnF_2)	releases both F^- and Sn^{+2} ions into the oral environment, which have cariostatic and antimicrobial properties, respectively; tooth staining and instability are the main disadvantages	dentifrices mouth rinses solutions gels prophylaxis pastes
Ammonium fluoride (NH_4F)	although investigated intensively some decades ago, it is currently unused – mainly due to its unpleasant taste and lack of superiority in clinical performance over NaF formulations	solutions
Titanium tetrafluoride (TiF_4)	able to significantly reduce enamel solubility (as solution), due to the formation of a glaze on enamel and dentine; currently being tested in solutions/varnishes as preventative for caries and erosion[1]	solutions varnishes
Organic compounds		
Amine fluoride	associated with a reduction in plaque adhesiveness due to the greater affinity of hydrophilic counter-ions to the enamel; also associated with complexed store of fluoride ions, which may enhance diffusion through carious enamel	dentifrices gels mouth rinses prophylaxis pastes
Silane fluoride	associated with complexed store of fluoride (similarly to amine fluoride); unlike NaF, MFP and SnF_2 (which dissolve in water and release fluoride ions), this compound is insoluble, releasing HF after contact with saliva, which diffuses into enamel more efficiently than fluoride ions	varnishes
Monofluorophosphate-containing compounds		
Sodium monofluorophosphate (Na_2FPO_3)	can be used in both neutral and acidic vehicles; fluoride is covalently bound in Na_2FPO_3 and requires hydrolysis in order to release fluoride ions; both F^- and PO_3F^{-2} ions can diffuse into plaque and enamel, mainly in acidic pH, but the role of PO_3F^{-2} ion is not well established; one of the main advantages is its compatibility with chalk-based abrasives	dentifrices (neutral pH) gels (neutral and acidic pH)

Sources: Ellwood et al. [2], Axelsson [7] and Pessan et al. [47].
[1] See Magalhães et al. [this vol., pp. 158–170].

phosphate fluoride (APF), AmF, NH$_4$F and TiF$_4$. Most of the studies described in the literature, however, used solutions of NaF, SnF$_2$ and APF. These have differed regarding the mode of application and frequency of use. NaF solutions were applied to all teeth and then allowed to dry for 3 min (4 weekly applications at 3, 7, 10 and 13 years of age), while SnF$_2$ and APF solutions had to be constantly applied to the teeth for 3 min (twice/year). Ripa [9] reviewed 35 clinical studies evaluating the effectiveness of fluoridated solutions, and concluded that NaF, SnF$_2$ and APF solutions presented similar clinical effectiveness (around 30%), despite differences in fluoride concentration, mode of application, frequency of use and other characteristics. With the introduction of fluoride gels, which are much easier and safer in clinical practice, fluoridated solutions are no longer used and, therefore, will not be described in detail in the present chapter [10].

Gels and Foams

In contrast to solutions, fluoride gels have been extensively used in both self-applied and professionally applied modalities, as gels are much more viscous than solutions. This property makes it possible to treat an entire arch at the same time, which reduces both the time of application and the risk of excessive ingestion of fluoride. When professionally used, gels can be applied with brushes or cotton pellets, but the use of trays (stock or custom made) minimizes the risk of excessive fluoride intake. Increased penetration of gels between the teeth can be achieved by using thixotropic products (which flow under pressure). If self-applied, gels are usually used in trays or with a toothbrush.

Due to the high fluoride concentration in gels (0.5% F$^-$ in SnF$_2$, 0.9% F$^-$ in neutral NaF and 1.23% F$^-$ in APF), care must be taken when using these products in order to avoid side effects. The patient must remain seated during application, suction devices must be used during application, and the patient must be instructed to spit out repeated times after application. Application time is 4 min for both neutral and acidic products [11].

There is consistent evidence for the benefits of gels in caries prevention [12, 13]. A Cochrane review included data from 23 trials (7,747 children and adolescents), in which fluoride gels were compared to a placebo treatment or no treatment (table 2) [14]. The effect of fluoridated gels in the permanent dentition was significantly influenced by the frequency of application, as well as the intensity of application (frequency × concentration) and self-application (which can be associated with a higher frequency of use). Therefore, fluoridated gels must be applied 2–4 times per year (table 3), depending on caries risk consideration, in order to achieve the expected benefits in caries prevention. No conclusion could be drawn for the deciduous dentition. The effect of fluoride gels obtained in the most recent meta-analysis (28% pooled DMFS prevented fraction) is not substantially different from that obtained more recently, when four new studies (published after the Cochrane review) were included [15].

Fluoridated foams became available more recently, with compositions similar to APF gels. However, as an amount of fluoride 4–5 times lower is used during application of foams (due to the reduced density of these products), they can be considered as a safer option regarding the risk of excessive fluoride intake. There are only a few studies comparing the effectiveness of foams and gels, and there seems to be little or no difference between clinical efficacy of both vehicles [16–18]. However, more studies with appropriate research protocols are still needed to address this issue before foams can be recommended as a substitute to gels.

Table 2. Results of the Cochrane reviews on the effectiveness of topical fluoride methods in caries prevention

Year of publi-cation	Method evaluated	Trials included in review and meta-analysis	Children included in meta-analysis	DMFS pooled PF with 95% CI %	dmfs pooled PF with 95% CI %	Main findings	Main conclusions
2002	gel	25 (23)	7,747	28 (19–37)	not available	effect varied according to type of control group used (DMFS PF 19% higher in non-placebo controlled trials); no significant association between DMFS PF and baseline caries severity or background fluoride exposure; effect influenced by increased frequency or intensity of application and self-application; only 2 trials reported on adverse events	clear evidence of a caries-inhibiting effect of fluoride gel; best estimate of the magnitude of this effect, based on the 14 placebo-controlled trials, is a 21% reduction (95% CI 14–28%) in DMFS PF
2002	varnish	9 (7)	2,709	46 (30–63)	33 (19–48)	no significant association between estimates of DMFS PF and baseline caries severity or background exposure to fluorides; power was limited due to the inclusion of few trials	substantial caries-inhibiting effect in both the permanent and the deciduous dentitions; little information concerning acceptability of treatment or possible side effects
2003	toothpaste	74 (70)	42,300	24 (2–28)	not available	effect increased with higher baseline levels of DMFS, higher fluoride concentration, higher frequency of use, and supervised brushing, but was not influenced by exposure to water fluoridation; little information concerning the deciduous dentition or adverse effects (fluorosis)	benefits of fluoride toothpastes in caries prevention firmly based on trials of relatively high quality

Table 2. Continued

Year of publi-cation	Method evaluated	Trials included in review and meta-analysis	Children included in meta-analysis	DMFS pooled PF with 95% CI %	dmfs pooled PF with 95% CI %	Main findings	Main conclusions
2003	mouthrinse	36 (34)	14,600	26 (23–30)	not available	no significant association between DMFS pooled PF and baseline caries severity, background exposure to fluorides, rinsing frequency and fluoride concentration	clear reduction in caries increment in children who regularly use fluoride mouth rinses at two main strengths (230 or 900 ppm) and rinsing frequencies (daily or weekly/fortnightly)

DMFS = Decayed/missing/filled surface (permanent teeth); dmfs = decayed/missing/filled surface (deciduous teeth); PF = prevented fraction: (mean increment in the control group – mean increment in the intervention group)/mean increment in the control group.

Varnishes

Fluoridated varnishes were introduced into the market in the 1960s, and are intended for professional application only. The main advantages of varnishes are the prolonged contact time between fluoride and the tooth surfaces (increases fluoride uptake by dental hard tissues, as well as the formation of CaF_2 reservoirs), and the possibility of using very small amounts of the product (a thin layer), which minimizes the risk of excessive fluoride ingestion. These products are much more concentrated than gels, with typical concentrations of 22,600 ppm fluoride (in NaF varnishes), 7,000 ppm fluoride (in difluorosilane varnishes) or 56,300 ppm fluoride (in 6% NaF + 6% CaF_2 varnishes) [19]. Duraphat (Inpharma, Germany – NaF) and Fluor Protector (Vivadent, Liechtenstein – difluorosilane) are the most used and studied products. In order to achieve the maximum benefits for caries prevention, varnishes must be applied 2–4 times/year (table 3), depending on caries-risk considerations.

There is evidence attesting to a substantial caries-inhibiting effect in both the permanent (46% reduction in pooled DMFS prevented fraction) and deciduous (33% reduction in pooled dmfs prevented fraction) dentitions, based on a Cochrane review including data from 2,709 children (table 2) [19]. As for rinses and gels, no significant association between estimates of DMFS prevented fractions and baseline caries severity or background exposure to fluorides was observed. The results of that systematic review, however, must be interpreted with caution, due to the low number of trials included, which limits the power of the analyses. A recent review found six trials not included in the Cocrhane review, and the pooled DMFS prevented fractions ranged from 34 to 57% [15].

Despite having higher fluoride concentrations, varnishes can be regarded as a safer option when compared to gels, due to the small amount used during application (table 4). Fluoride concentrations in plasma and urine of children were reported to be lower than toxic levels after the application of a fluoride varnish [20, 21].

Table 3. Evidence-based clinical recommendations for professionally applied topical fluoride

| | Age category of recall patients | | | | | | | | |
| | <6 years | | | 6–18 years | | | >18 years | | |
	recommendation	grade of evidence	strength of recom-mendation	recommendation	grade of evidence	strength of recom-mendation	recommendation	grade of evidence	strength of recom-mendation
Low risk	may not receive additional benefit from professional topical fluoride application[1]	Ia	B	may not receive additional benefit from professional topical fluoride application[1]	Ia	B	may not receive additional benefit from professional topical fluoride application[1]	IV	D
Mode-rate risk	varnish application at 6-month intervals	Ia	A	varnish application at 6-month intervals	Ia	A	varnish application at 6-month intervals	IV	D[2]
				OR			OR		
				fluoride gel application at 6-month intervals	Ia	A	fluoride gel application at 6-month intervals	IV	D[3]
High risk	varnish application at 6-month intervals	Ia	A	varnish application at 6-month intervals	Ia	A	varnish application at 6-month intervals	IV	D[2]
	OR			OR			OR		
	varnish application at 3-month intervals	Ia	D[4]	varnish application at 3-month intervals	Ia	A	varnish application at 3-month intervals	IV	D[2]
				OR			OR		
				fluoride gel application at 6-month intervals	Ia	A	fluoride gel application at 6-month intervals	IV	D[3]
				OR			OR		
				fluoride gel application at 3-month intervals	IV	D[3]	fluoride gel application at 3-month intervals	IV	D[3]

Evidence from systematic reviews of randomized controlled trials (Ia) or expert committee reports or opinions or clinical experience of respected authorities (IV).

Laboratory data have demonstrated the equivalence of foam to gels in terms of fluoride release; however, only 2 clinical trials have been published evaluating its effectiveness. Because of this, the recommendations for use of fluoride varnish and gels have not been extrapolated to foams.

Because there is insufficient evidence to address whether or not there is a difference in the efficacy of sodium fluoride vs. acidulated phosphate fluoride gels, the clinical recommendations do not distinguish between these formulations. Application time for fluoride gel and foam should be 4 min. A 1-min fluoride application is not endorsed.

Source: American Dental Association [56].

[1] Fluoridated water and toothpastes may already provide adequate caries prevention. Decisions on whether to apply topical fluoride should balance this consideration with professional judgment and patient preferences.

[2] Although there are no clinical trials, there is reason to believe that fluoride varnish would work similarly in this age group.

[3] Although there are no clinical trials, there is reason to believe that fluoride gels would work similarly in this age group.

[4] Emerging evidence indicates that applications more frequent than twice per year may be more effective in preventing caries.

Table 4. Comparison of professionally applied fluoride methods (varnish, gel and foam) regarding their effectiveness, clinical use, toxicity, cost and patient acceptance.

	Caries prevention	Clinical application	Fluoride ingestion	Cost	Acceptability
Varnish	effective in high-risk children (permanent teeth)	easy; application time varies	lowest risk; moisture can be better controlled than gel or foam	most expensive	preferred by patients and hygienists over gel
Gel	effective in high-risk children (permanent teeth)	easy; 4-min application time	% retained can be substantial; procedure must be followed to reduce risk	low cost	well-tolerated by most patients, but varnish is preferred
Foam	not clinically tested; likely to be similar to gel	easy; 4-min application time	risk of over-ingestion is less compared with gel	low cost	not formally assessed; likely to be to similar to gel

Source: Hawkins et al. [59] (modified).

Rinses

Fluoride mouth rinses have been successfully used in dentistry for about 6 decades, either as self-application or community-based methods. NaF solutions are the most widely used, although formulations containing other fluoride compounds are also available. Typically, solutions containing 230 ppm fluoride are intended for daily use at home, while a higher concentration (900 ppm fluoride) is used in community-based programs at weekly/fortnightly intervals. The main advantages of the method include effectiveness, simplicity of use and the possibility of application by a non-dental professional, which ultimately affects cost.

A clear reduction (26% in pooled DMFS prevented fraction) in caries increments in the permanent dentition of children who regularly use fluoride rinses was found by a recent meta-analysis of 34 clinical trials (involving 14,600 children; table 2) [22]. Such findings apply to both daily and weekly/fortnightly rinses with 230- and 900-ppm fluoride solutions, respectively, indicating that the mode of application will depend on personal preferences (when used at home) and on the availability of personnel to supervise the use of the solutions (in school-based programs). No significant association was found between DMFS pooled prevented fraction and baseline caries severity, background exposure to fluorides, rinsing frequency and fluoride concentration. No conclusion could be drawn for the deciduous dentition. The benefits of mouthrinsing were shown to be affected by subsequent discontinuation [23].

Dentifrices

Fluoridated dentifrices are by far the most widespread form of fluoride delivery and are currently used by over 500 million people worldwide [24].

Table 5. Ingredients commonly used in fluoridated dentifrices

Active agents	Other compounds
fluoride – 1 compound or 2 (in combination)	abrasive particles
agents for enhancement of the fluoride effect	detergents, foaming agents
chemical plaque control agents	flavoring agents, preservatives, and coloring agents
anti-calculus agents	thickeners, agents to regulate viscosity
buffering systems	water

Source: Axelsson [7] (modified).

Considering the multifactorial etiology of dental caries, toothbrushing with a fluoridated dentifrice can be regarded as the best method of fluoride use, as it combines the mechanical removal or disruption of dental plaque (which is also beneficial in periodontal health maintenance) with the caries-protective effect of fluoride [25]. It has been regarded as the method of choice by public health authorities, as it is convenient, inexpensive, culturally accepted and widespread [26]. Besides the therapeutic properties of dentifrices, their cosmetic benefits (related to cleanliness, removal of stains, whiteness and protection against oral malodor) constitute additional reasons for the wide acceptance of this method [2].

Dentifrices are available as various formulations of gels and pastes, and may contain bleaching, anti-plaque and desensitizing ingredients, with labeling and flavoring characteristics directed to adults and children. They must not be confounded with prophylaxis pastes (which have higher fluoride content, are more abrasive and are used less frequently) or gels (which do not have abrasive particles, are much more concentrated and used at a lower frequency) [4]

Composition

As previously mentioned, the greatest advantages of dentifrices are the removal or disruption of dental plaque associated with fluoride delivery to dental hard tissues and intraoral reservoirs (especially dental plaque). In order to achieve these goals, the various formulations available may contain the components listed in table 5. Dentifrices can vary widely in their composition, depending on the benefits that each formulation intends to provide (i.e. anti-plaque, anti-calculus, whitening). Due to the scope of this chapter, however, only factors related to the anti-caries properties of fluoridated dentifrices will be discussed.

The abrasive system is an important component that can affect fluoride availability, which will ultimately interfere with its clinical performance. It is known that the first formulations of toothpastes failed to show a significant effect on the reduction in caries levels, due to the use of incompatible fluoride compounds and abrasive systems [27]. The compatibility of MFP, NaF and other formulations with the abrasive systems most commonly used are [2, 7, 28]:

- MFP:
  Alumina trihydrate ($Al_2O_3 \cdot 3H_2O$)
  Anhydrous dicalcium phosphate ($CaHPO_4$)
  Dicalcium phosphate dihydrate ($CaHPO_4 \cdot 2H_2O$)
  Calcium carbonate ($CaCO_3$)
- MFP, NaF and other formulations
  Calcium phosphate ($Ca_2P_2O_7$)
  Hydrated silica (SiO_2)
  Sodium bicarbonate ($NaHCO_3$)
  Insoluble sodium metaphosphate ($NaPO_3)_x$
  Acrylic polymer

Clinical Efficacy

Marinho et al. [4] evaluated the effect of fluoridated dentifrices in caries increments through a meta-analysis involving data from 42,300 children and adolescents (table 2). Due to the large number of trials (n = 70) of relatively high quality, there is strong evidence for the clinical effectiveness of toothpastes in preventing caries in that age group (pooled DMFS prevented fraction was 24%). Unlike for rinses, gels and varnishes, the effect of fluoridated toothpastes increased with higher baseline levels of DMFS, higher fluoride concentration, higher frequency of use and supervised brushing. Background exposure to fluoridated water did not influence the effects of toothpastes in caries reduction. Even considering the large number of trials and subjects evaluated, little information was found concerning the deciduous dentition or adverse effects, such as fluorosis. After the publication of that Cochrane review [4], ten new studies addressing the same topic were published (between 2002 and 2008) and the pooled DMFS prevented fraction was around 25% [29], not substantially different from that found by Marinho et al. [4].

Factors Affecting Clinical Efficacy

Among the factors that affect the clinical efficacy of toothpaste formulations, fluoride concentration and pH are the most relevant and, therefore, special emphasis will be given to these. Other factors include frequency of brushing, amount of toothpaste used, rinsing behavior after brushing, timing and duration of brushing, the fluoride compound in the toothpaste and age when brushing commenced. These will be briefly summarized.

Fluoride Concentration

It has been suggested that the fluoride concentration of toothpastes is one of the main determinants of their efficacy. Results of large clinical trials comparing the effectiveness of toothpastes in the range of 1,000–2,500 ppm fluoride indicate that for every 500-ppm increase in toothpaste fluoride concentration, an additional 6% reduction in caries is obtained [27, 30]. As it could be wrongly assumed from those results that toothpastes with higher fluoride concentrations should always be preferred in order to achieve the maximum benefits of these products, this topic needs to be further explored, mainly in terms of risks/benefits.

Typically, low-fluoride toothpastes (usually containing 500–550 ppm) have been recommended to children under 7 years of age, in order to minimize fluoride ingestion from this source. Toothpastes with fluoride concentrations in the range of 1,000–1,500 ppm are usually indicated for children older than 7 years, adolescents and adults, and higher concentrations would be indicated for patients at high caries risk or to prevent root caries [26, 31].

The results of a recent systematic review of studies comparing the clinical efficacy of toothpastes with different fluoride concentrations are summarized in table 6. When compared to placebo, significant differences in caries increments were only seen for formulations containing 1,000 ppm fluoride or above [6], suggesting that low-fluoride toothpastes are less effective for caries control when compared to conventional formulations. Care must be taken when interpreting these results, in order to avoid misleading the reader.

First of all, the conclusion that the efficacy of dentifrices containing 450–550 ppm is not significantly different from placebo was based only on 2 trials, while the number of studies comparing placebo with conventional formulations (1,000–1,500 ppm) was substantially higher (58 trials). In addition, no conclusion could be taken when comparing the clinical efficacy of low-fluoride and conventional toothpastes, as only one trial met the inclusion criteria of that review, clearly indicating that additional randomized controlled trials are still needed to fully address this issue.

There is reason to be believe that the differences between low-fluoride and conventional toothpastes may not be as large as might be expected.

Table 6. Comparison of the clinical effectiveness of toothpastes with different fluoride concentrations

Comparisons	Studies included in the meta-analysis, n	Pooled DMFS prevented fraction with 95% CI	p value
Placebo vs.			
250 ppm	3	8.9 (−1.6 to 19.4)	0.097
440/500/550 ppm	2	7.9 (−6.1 to 21.9)	0.27
1,000/1,055/1,100/1,250 ppm	54	22.2 (18.7 to 25.7)	<0.00001
1,450/1,500 ppm	4	23.0 (15.3 to 28.9)	<0.00001
2,400/2,500/2,800 ppm	4	36.6 (17.5 to 55.6)	<0.00001
250 ppm vs.			
1,000/1,055/1,100/1,250 ppm	2	16.8 (8.5 to 25.1)	0.000076
440/500/550 ppm vs.			
1,000/1,055/1,100/1,250 ppm	1	0.5 (−15.0 to 16.0)	n.c.
2,400/2,500/2,800 ppm	1	12.7 (−1.7 to 27.0)	n.c.
1,450/1,500 ppm	6	9.6 (2.5 to 16.6)	0.0078
1,000/1,055/1,100/1,250 ppm vs.			
1,700/2,000/2,200 ppm	2	9.4 (2.1 to 16.8)	0.011
2,400/2,500/2,800 ppm	6	12.2 (6.0 to 18.4)	0.00012

Source: Walsh et al. [6] (modified). n.c. = p value not calculated, as there was only 1 study for each comparison.

A recent randomized clinical trial demonstrated that the clinical performance of low-fluoride toothpastes is dependent on caries activity. It was shown that the clinical efficacy of a 500-ppm fluoride dentifrice was similar to that of a 1,100-ppm fluoride dentifrice when used by caries-inactive children, but the low-fluoride dentifrice was less effective than the conventional formulation in controlling the progression of lesions in caries-active children [32]. More recently, it has also been demonstrated that plaque fluoride concentrations (an indirect indicator of clinical efficacy of topical fluoride products [33]), observed 1 h after brushing with conventional and low-fluoride toothpastes, were not significantly different between children residing in communities with fluoridated and non-fluoridated drinking water [34]. Finally, for the deciduous dentition, uncertainty regarding the effectiveness of low-fluoride toothpastes for preventing caries was reported due to the lack of trials [6].

Considering the evidence above, along with the conclusion of the systematic review by Walsh et al. [6] on the uncertainty surrounding the estimates of dentifrices containing 450–550 ppm fluoride, it becomes clear that caution must be taken when recommending low-fluoride toothpastes for the prevention of caries in the deciduous dentition. While no conclusive evidence is available, the decision of recommending those formulations will depend on caries activity, besides professional and patient choices regarding risks/benefits. It seems reasonable to recommend low-fluoride (500 ppm) toothpastes for young children who are at risk of developing fluorosis in the permanent maxillary central incisors (less than 3 years of age) but at low caries risk, especially if they live in a fluoridated area. In all other

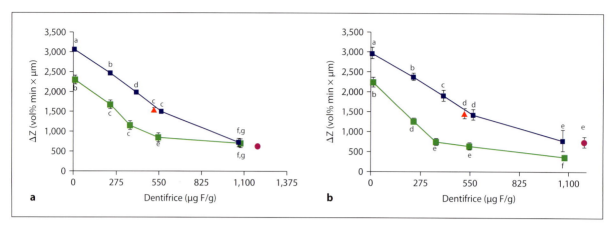

Fig. 1. Mean mineral loss (ΔZ) for the different fluoride concentrations in dentifrices at pH 5.5 (**a**) [35] and pH 4.5 (**b**) [36], compared to neutral formulations. Blue squares = Neutral toothpastes; Green squares = Acidic toothpastes; Triangle = Commercial children's toothpaste; Circle = Commercial 1,100 µg/g toothpaste. Means followed by distinct letters are significantly different (Kruskal-Wallis, p<0.05). Bars indicate standard deviations.

cases, toothpastes containing at least 1,000 ppm fluoride should be used.

pH of the Dentifrice

The controversial evidence on the use of low-fluoride dentifrices for caries control has led to increasing interest in strategies to improve the effectiveness of such formulations. Due to the inverse relationship between CaF_2 formation and pH, dentifrices with acidic pH have been tested in clinical and laboratory studies. Results from in vitro experiments showed that the effectiveness of dentifrices containing 550 ppm fluoride (pH 5.5) [35] or 412 ppm fluoride (pH 4.5) [36] were similar to that of a 1,100-ppm dentifrice at neutral pH (fig. 1). These results were confirmed in a recent clinical trial evaluating caries progression in the deciduous dentition of high caries-risk children living in a fluoridated area. Caries progression rates observed 20 months after the use of a 550-ppm dentifrice in pH 4.5 were similar to those seen for the conventional 1,100-ppm neutral toothpaste [37]. The superior clinical performance of the acidic toothpastes can be partially explained by an increased plaque fluoride uptake

when compared with neutral formulations [38]. Formulations with a pH of 4.5 were also shown to have similar abrasiveness in comparison with neutral dentifrices with the same fluoride concentrations [39].

Fluoride Compound Used in the Formulation

Another factor that has been traditionally associated with the clinical efficacy of dentifrices is the active fluoride agent used in the formulation (table 1). Most marketed dentifrices contain sodium fluoride (NaF), or sodium monofluorophosphate (SMFP), although formulations with SnF_2 and amine fluoride are also available in some countries [2, 40]. There used to be a controversy regarding the clinical efficacy obtained by NaF and SMFP toothpastes (the most widely used formulations). Based on the premise that fluoride only exerts its effects on de- and remineralization as a free ion, several reports have claimed the superiority of NaF formulations (which releases free F^-), in comparison with SMFP, where fluoride is covalently bound to phosphate, and requires enzymatic hydrolysis to release free F^-.

Table 7. Potential factors affecting clinical effectiveness of toothpastes

	Interpretation or recommendation	Evidence
Fluoride concentration	increase in prevented fraction (permanent dentition) for 1,000 ppm fluoride	1
Rinsing behaviour	discourage rinsing with large volumes of water; encourage young children to spit out excess toothpaste	4, 5
Frequency of brushing	increase in prevented fraction moving from once to twice a day	1
Supervision	lower prevented fraction with unsupervised toothbrushing	1
When to brush	brush last thing at night and on one other occasion	4, 5
Type of fluoride	toothpastes containing sodium fluoride, sodium monofluorophosphate or stannous fluoride are clinically effective	1, 5
Age to commence brushing	advise parents/carers to begin brushing once the primary teeth have commenced eruption	4, 5
Background fluorides	higher prevented fraction in presence of any background fluoride (e.g. in water)	1
Mean initial caries	increase in prevented fraction per unit increase in mean initial level of caries	1

Evidence levels: 1 = systematic review of at least one randomized controlled trial; 2 = at least 1 randomized controlled study; 3 = non-randomized intervention studies; 4 = observational studies; 5 = traditional reviews, expert opinion. Sources: Marinho [5] and Davies et al. [58].

The clinical efficacy of different formulations of toothpastes has been addressed by several clinical studies and literature reviews over the last decades with differing conclusions [4, 40–45]. Reviews conducted in the 1990s claim a difference in effectiveness of 6–7% favoring NaF formulations [42, 43]. A recent Cochrane review comparing formulations containing SMFP (22 trials), SnF$_2$ (19 trials), NaF (10 trials) and amine fluoride (5 trials) did not find an association between the main types of fluoride compounds present in toothpaste formulations and the magnitude of the treatment effect [4]. Nevertheless, the authors of that review considered their result to be less reliable than evidence from head-to-head comparisons. The long-term significance of this difference has been the subject of debate, so the question remains to be further investigated.

Other Factors Affecting Clinical Efficacy

As for all other vehicles of fluoride delivery, the primary goal is to enrich the intraoral fluoride reservoirs at levels that can interfere with the dynamics of dental caries. Therefore, it is reasonable to assume that toothbrushing habits able to increase or sustain intraoral fluoride levels will have a positive effect on the clinical efficacy of toothpastes. Other aspects, including mean initial caries and background exposure to fluorides, can also affect the clinical outcome when using a fluoridated dentifrice. However, the degree of scientific evidence surrounding these assumptions varies considerably among the factors that may influence the clinical efficacy of toothpastes. Some factors that potentially influence the effectiveness of fluoride dentifrices are summarized in table 7.

Slow-Release Fluoride Devices

The rationale for the use of slow-release fluoride devices is that salivary fluoride concentrations are significantly increased during the entire day without relying on patient compliance [46]. Two main types of devices have been currently used – the copolymer membrane (developed in the USA) and the glass beads (developed in the UK) – although a third type (which contains a mixture of NaF and hydroxyapatite) has recently been introduced [47].

The copolymer membrane device was designed as a membrane-controlled reservoir-type, having an inner core of a 50:50 mixture of hydroxyethyl methacrylate (HEMA)/methyl methacrylate (MMA) copolymer (which contains NaF), surrounded by a 30:70 HEMA/MMA copolymer membrane (which controls the rate of fluoride release from the device) [48]. It is usually attached to the buccal surface of the first permanent molar by means of stainless steel retainers or bonded to the tooth surfaces using adhesive resins. Salivary fluoride levels were shown to remain significantly elevated throughout a 100-day test period when using this type of device [49, 50].

The glass device dissolves slowly when moist in saliva, releasing fluoride without significantly affecting the device's integrity. The original device was dome shaped [46, 51], and usually attached to the buccal surface of the first permanent molar using adhesive resin. Due to low retention rates, it was later substantially changed to a kidney-shaped device, and more recently to the form of a disk that is placed within a plastic bracket, the latter substantially improving device handling, attachment and replacement (without the need for de-bonding). In contrast to the copolymer membrane device, the glass type has been shown to have a better longevity, releasing fluoride continuously for up to 2 years [51].

The hydroxyapatite-Eudragit RS100 diffusion-controlled fluoride system is the newest type of slow-release device and is intended to release 0.15 mg/day. This device was shown to promote a significant increase in salivary and urinary fluoride concentrations for at least 1 month, but to date there has been only one trial evaluating this kind of device [52].

Studies in humans and animals attest to the safety of these devices regarding toxicity if a device is swallowed [46, 53]. The copolymer device was shown to be clinically effective in reducing caries incidence in rats by 63% [49], and in reducing dentine sensitivity in humans 4 weeks after its use in patients under periodontal therapy [54]. In children, the only randomized controlled trial showed significantly lower caries increments in subjects using fluoride glass beads when compared to placebo glass beads, both for DMFS (0.84 and 2.34, $p < 0.05$) and dmfs (2.26 and 8.41, $p < 0.001$) [51]. Retention was the main problem associated with the use of the devices.

The fluoride slow-release glass devices have a number of important potential applications in addition to their use for caries prevention in high-caries-risk groups and a number of current research studies are investigating their efficacy. A randomized double-blind study investigating the effect of the slow-release glass devices to prevent demineralization around orthodontic brackets is currently being conducted in Leeds, UK. Preliminary in situ studies have also shown that these devices are beneficial for the prevention of dentinal root caries, which is an increasing problem for the aged population. The results of these clinical studies are eagerly awaited.

Combinations of Topical Fluoride Modalities

Given the nature of the various modalities of fluoride delivery (community, self-applied and professional methods), it is not uncommon to observe that individuals are frequently exposed to two or more methods. Based on the mechanism of action of fluoride, it is reasonable to assume that the use of different fluoride vehicles would lead

Table 8. DMFS (pooled) estimates of treatment effects (as prevented fraction) for direct comparisons between fluoride gels, varnishes, rinses, and toothpastes

TFT types	Number of studies	Prevented fraction, %	95% CI, %
Varnish vs. gel	1	14	−12 to 40
Varnish vs. mouth rinse	4	10	−12 to 32
Gel vs. mouth rinse	1	−14	−40 to 12
Toothpaste vs. gel	3	0	−21 to 21
Toothpaste vs. mouth rinse	6	0	−18 to 19
Toothpaste vs. any TFT[1]		1	−13 to 14
Toothpaste + varnish vs. toothpaste alone	1	48	12 to 84
Toothpaste + gel vs. toothpaste alone	3	14	−9 to 38
Toothpaste + mouth rinse vs. toothpaste alone	5	7	0 to 13
Toothpaste + any TFT vs. toothpaste alone	9	10	2 to 17

TFT = Topical fluoride treatment. Source: Marinho [5].
[1] Gel trials (n = 3) and mouth rinse trials (n = 6), but no varnish trial.

to enhanced preventive effects when compared to a single vehicle alone. However, the magnitude of the benefits of topical fluoride treatments used together, as well as the possible side effects, has been recently questioned.

This particular issue was addressed in a systematic review assessing the effectiveness of a combination of topical fluoride methods versus a single topical fluoride method. The DMFS pooled prevented fraction was only 10% in favor of the combined regimens (mouth rinses, gels or varnishes used in combination with toothpaste) when compared to dentifrice alone, but most of the individual comparisons were not statistically significant [55]. Those results, however, must be interpreted with caution, as it may be prematurely concluded that the association of two modalities of fluoride delivery is not beneficial. The meta-analysis involved a low number of trials (a small number of trials was included in each relevant comparison) and, according to the authors, the review has not tested all combinations of possible practical value. The main results from the comparisons involving

use of toothpaste alone or combined with different fluoride modalities are listed in table 8.

Even considering that topical fluorides used in addition to fluoride toothpaste achieve a modest reduction in caries compared to toothpaste used alone [55], it is still recommended that patients at high caries risk receive additional fluoride therapy. For patients at low caries risk, however, additional fluoride therapy may be not only ineffective, but may also increase the risk of side effects. Therefore, the decision of using additional fluoride therapy must be based in both cost/benefit and risk/benefit considerations. The recommendations for topical fluoride therapy from the American Association Council on Scientific Affairs [56], according to the caries risk, are listed in table 3.

Concluding Remarks

The caries-protective benefits of the various modalities of fluoride therapy have been confirmed by over 60 years of clinical and laboratory studies.

There is strong evidence that the use of fluoridated toothpastes, gels, varnishes and mouth rinses is effective in controlling the progression of carious lesions. Also, there is evidence that additional reductions in dental caries can be achieved by combining the use of a fluoride toothpaste with another form of topical fluoride, although the size of the caries-preventive effect may not be substantial [57]. In this case, the combination of different fluoride vehicles may not be justifiable for young children, due to the possibility of increasing fluoride intake, which ultimately might increase the risk of dental fluorosis. On the other hand, any extra benefit could have a great impact on the oral health of high-caries-risk individuals.

The publication of systematic reviews made a considerable contribution to the understanding of the real benefits of fluoride therapy in caries control. Despite the substantial evidence that currently exists, there is still a need for future randomized controlled trials with adequate protocols to establish the actual benefits of each mode of fluoride application, so both clinicians and patients can make a better decision regarding the most appropriate form of use for fluoride. Presently, considering that no modality of topical fluoride administration (mouth rinses, gels, varnishes and dentifrices) has been proven to be more substantially effective than any other [54], the advantages of dentifrices – regarding cost, availability, cosmetic benefits, and impact on both caries and periodontal health – indicate that this should be the fluoride vehicle of choice. Bearing in mind the different fluoride agents and products with varying concentrations of fluoride that have been scientifically proven to be effective against dental caries, it can sometimes be confusing to the patient and challenging to the dental clinician to make the most appropriate choice. Professional dental academies and national societies – e.g. The American Dental Association (ww.ada.org) as well as both the American (www.aapd.org) and European Academies of Paediatric Dentistry (www.eapd.eu) – produce guidelines to help dental practitioners and the general public understand the appropriate use of fluoride products.

References

1 Featherstone JDB: Prevention and reversal of dental caries: role of low level fluoride. Community Dent Oral Epidemiol 1999;27:31–40.

2 Ellwood RP, Fejerskove O, Cury JA, Calrkson B: Fluoride in caries control; in Fejerskov O, Kidd EAM (eds): Dental Caries: The Disease and Its Clinical Management, ed 2. Oxford, Blackwell & Munksgaard, 2008, p 287–323.

3 Marinho VC, Higgins JP, Logan S, Sheiham A: Topical fluoride (toothpastes, mouthrinses, gels or varnishes) for preventing dental caries in children and adolescents. Cochrane Database Syst Rev 2003;4:CD002782.

4 Marinho VC, Higgins JP, Sheiham A, Logan S: Fluoride toothpastes for preventing dental caries in children and adolescents. Cochrane Database Syst Rev 2003;1:CD002278.

5 Marinho VCC: Cochrane reviews of randomized trials of fluoride therapies for preventing dental caries. Eur Arch Paediatr Dent 2009;10:183–191.

6 Walsh T, Worthington HV, Glenny AM, Appelbe P, Marinho VC, Shi X: Fluoride toothpastes of different concentrations for preventing dental caries in children and adolescents. Cochrane Database Syst Rev 2010;1:CD007868.

7 Axelsson P: Use of fluorides; in Axelsson P (ed): Preventive Materials, Methods and Programs. Slovakia, Quintessence Books, 2004, p 263–268.

8 Newbrun E: Evolution of professionally applied topical fluoride therapies. Compend Contin Educ Dent 1999;20(suppl 1):5–9.

9 Ripa LW: Professionally (operator) applied topical fluoride therapy: a critique. Int Dent J 1981;31:105–120.

10 Horowitz HS, Ismail AI: Topical fluorides in caries prevention; in Fejerskov O, Ekstrand J, Burt BA (eds): Fluoride in Dentistry. Copenhagen, Munksgaard Textbook, 1996, p 311–327.

11 Hawkins R, Locker D, Noble J, Kay EJ: Prevention. Part 7: professionally applied topical fluorides for caries prevention. Br Dent J 2003;195:313–317.

12 Ripa LW: Review of the anticaries effectiveness of professionally applied and self-applied topical fluoride gels. J Public Health Dent 1989;49:297–309.

13 van Rijkom HM, Truin GJ, van't Hof MA: A meta-analysis of clinical studies on the caries-inhibiting effect of fluoride gel treatment. Caries Res 1998;32:83–92.

14 Marinho VC, Higgins JP, Logan S, Sheiham A: Fluoride gels for preventing dental caries in children and adolescents. Cochrane Database Syst Rev 2002; 2:CD002280.

15 Poulsen S: Fluoride containing gels, mouthrinses and varnishes: an update of efficacy. Eur Arch Paediatr Dent 2009; 10:157–161.

16 Jiang H, Tai B, Du M, Peng B: Effect of professional application of APF foam on caries reduction in permanent first molars in 6–7-year-old children: 24-month clinical trial. J Dent 2005; 33469–33473.

17 Wei SHY, Chik FF: Fluoride retention following topical fluoride foam and gel application. Pediatr Dent 1990;12: 368–374.

18 Whitford GM, Adair SM, Hanes CM, Perdue EC, Russel CM: Enamel uptake and patient exposure to fluoride: comparison of APF gel and foam. Pediatr Dent 1995;17:199–203.

19 Marinho VC, Higgins JP, Logan S, Sheiham A: Fluoride varnishes for preventing dental caries in children and adolescents. Cochrane Database Syst Rev 2002;3:CD002279.

20 Ekstrand J, Koch G, Petersson LG: Plasma fluoride concentration and urinary fluoride excretion in children following application of the fluoride-containing varnish Duraphat. Caries Res 1980;14:185–189.

21 Pessan JP, Pin MLG, Martinhon CCR, da Silva SMB, Granjeiro JM, Buzalaf MAR: Analysis of fingernails and urine as biomarkers of fluoride exposure from dentifrice and varnish in 4- to 7-year-old children. Caries Res 2005;39:363–370.

22 Marinho VC, Higgins JP, Logan S, Sheiham A: Fluoride mouthrinses for preventing dental caries in children and adolescents. Cochrane Database Syst Rev 2003;3:CD002284.

23 Yamaguchi N, Saito T, Oho T, Sumi Y, Yamashita Y, Koga T: Influence of the discontinuation of a school-based, supervised fluoride mouthrinsing programme on the prevalence of dental caries. Community Dent Health 1997;14: 258–61.

24 World Health Organization: Oral Health. www.who.int/oral_health/media/en/index1.html (accessed 9 December 2010).

25 Pessan JP, Sicca CM, de Souza TS, da Silva SMB, Whitford GM, Buzalaf MAR: Fluoride concentrations in dental plaque and saliva after the use of a fluoride dentifrice preceded by a calcium lactate rinse. Eur J Oral Sci 2006;114:489–493.

26 Toumba KJ: Guidelines on the use of fluoride in children: an EAPD policy document. Eur Arch Paediatr Dent 2009; 10:129–135.

27 Stephen KW, Creanor SL, Russell JI, Burchell CK, Huntington E, Downie CF: A 3-year oral health dose-response study of sodium monofluorophosphate dentifrices with and without zinc citrate: anticaries results. Community Dent Oral Epidemiol 1988;16:321–325.

28 Richards A, Banting DW: Fluoride toothpastes; in Fejerskov O, Ekstrand J, Burt BA (eds): Fluoride in Dentistry. Copenhagen, Munksgaard Textbook, 1996, p 328–346.

29 Twetman S: Caries-prevention with fluoride toothpaste in children – an update. Eur Arch Paediatr Dent 2009;10:162–167.

30 O'Mullane DM, Kavanagh D, Ellwood RP, Chesters RK, Schafer F, Huntington E, Jones PR: A three-year clinical trial of a combination of trimetaphosphate and sodium fluoride in silica toothpastes. J Dent Res 1997;76:1776–1781.

31 Nordström A, Birkhed D. Fluoride retention in proximal plaque and saliva using two NaF dentifrices containing 5,000 and 1,450 ppm F with and without water rinsing. Caries Res 2009;43:64–69.

32 Lima TJ, Ribeiro CC, Tenuta LMA, Cury JA: Low-fluoride dentifrice and caries lesion control in children with different caries experience: a randomized clinical trial. Caries Res 2008;42:46–50.

33 Duckworth RM, Morgan SN, Gilbert RJ: Oral fluoride measurements for estimation of the anti-caries efficacy of fluoride treatments. J Dent Res 1992;(71Spec No):836–840.

34 Pessan JP, Alves KMRP, Ramires I, Taga MFL, Sampaio FC, Whitford GM, Buzalaf MAR: Effects of regular and low-fluoride dentifrices on plaque fluoride. J Dent Res 2010;89:1106–1110.

35 Brighenti FL, Delbem ACB, Buzalaf MAR, Oliveira FAL, Ribeiro DB, Sassaki KT: In vitro evaluation of acidified toothpastes with low fluoride content. Caries Res 2006;40:239–244.

36 Alves KMRP, Pessan JP, Brighenti FL, Franco KS, Oliveira FAL, Buzalaf MAR, Sassaki KT, Delbem ACB: In vitro evaluation of the effectiveness of acidic fluoride dentifrices. Caries Res 2007;41: 263–267.

37 Vilhena FV, Olympio KPK, Lauris JRP, Delbem ACB, Buzalaf MAR: Low-fluoride acidic dentifrice: a randomized clinical trial in a fluoridated area. Caries Res 2010;44:478–484.

38 Buzalaf MAR, Vilhena FV, Iano FG, Grizzo L, Pessan JP, Sampaio FC, Oliveira RC: The effect of different fluoride concentrations and pH of dentifrices on plaque and nail fluoride levels in young children. Caries Res 2009;43: 147–154.

39 Alves KM, Pessan JP, Buzalaf MA, Delbem AC: Short communication: in vitro evaluation of the abrasiveness of acidic dentifrices. Eur Arch Paediatr Dent 2009;10(suppl 1):43–45.

40 Volpe AR, Petrone ME, Davies R, Proskin HM: Clinical anticaries efficacy of NaF and SMFP dentifrices: overview and resolution of the scientific controversy. J Clin Dent 1995;(6 Spec No): 1–28.

41 Chaves SC, Vieira-da-Silva LM: Anticaries effectiveness of fluoride toothpaste: a meta-analysis. Rev Saude Publica 2002;36:598–606.

42 Stamm JW: Clinical studies of neutral sodium fluoride and sodium monofluorophosphate dentifrices; in Bowen WH (eds): Relative Efficacy of Sodium Fluoride and Sodium Monofluorophosphate as Anti-Caries Agents in Dentifrices. London, Royal Society of Medicine, 1995, pp 43–58.

43 Johnson MF: Comparative efficacy of NaF and SMFP dentifrices in caries prevention: a meta-analytic overview. Caries Res 1993;27:328–336.

44 Stookey GK, DePaola PF, Featherstone JD, Fejerskov O, Möller IJ, Rotberg S, Stephen KW, Wefel JS: A critical review of the relative anticaries efficacy of sodium fluoride and sodium monofluorophosphate dentifrices. Caries Res 1993;27:337–360.

45 Kingman A: Methods of projecting long-term relative efficacy of products exhibiting small short-term efficacy. Caries Res 1993;27:322–327.

46 Toumba KJ: Slow-release devices for fluoride delivery to high-risk individuals. Caries Res 2001;35(suppl 1):10–13.

47 Pessan JP, Al-Ibrahim NS, Buzalaf MA, Toumba KJ: Slow-release fluoride devices: a literature review. J Appl Oral Sci 2008;16:238–246.

48 Mirth DB: The use of controlled and sustained release agents in dentistry: a review of applications for the control of dental caries. Pharmacol Ther Dent 1980;5:59–67.

49 Mirth DB, Adderly DD, Amsbaugh SM, Monell-Torrens E, Li SH, Bowen WH: Inhibition of experimental dental caries using an intraoral fluoride-releasing device. J Am Dent Assoc 1983;107:55–58.

50 Mirth DB, Shern RJ, Emilson CG. Adderly DD, Li SH, Gomez IM: Clinical evaluation of an intra-oral device for the controlled release of fluoride. J Am Dent Assoc 1982;105:791–797.

51 Toumba KJ, Curzon MEJ: A clinical trial of a slow-releasing fluoride device in children. Caries Res 2005;39:195–200.

52 Altinova YB, Alaçan A, Aydin A, Sanisoglu SY: Evaluation of a new intraoral controlled fluoride release device. Caries Res 2005;39:191–194.

53 Curzon MEJ, Toumba KJ: In vitro and in vivo assessment of a glass slow fluoride releasing device: a pilot study. Brit Dent J 2004;196:543–546.

54 Marini I, Checchi L, Vecchiet F, Spiazzi L: Intraoral fluoride releasing device: a new clinical therapy for dentine sensitivity. J Periodontol 2000;71:90–95.

55 Marinho VC, Higgins JP, Sheiham A, Logan S: Combinations of topical fluoride (toothpastes, mouthrinses, gels, varnishes) versus single topical fluoride for preventing dental caries in children and adolescents. Cochrane Database Syst Rev 2004;1:CD002781.

56 American Dental Association Council on Scientific Affairs: Professionally applied topical fluoride – evidence-based clinical recommendations. J Am Dent Assoc 2006;137:1151–1159.

57 Salanti G, Marinho V, Higgins JP: A case study of multiple-treatments meta-analysis demonstrates that covariates should be considered. J Clin Epidemiol 2009;62:857–864.

58 Davies RM, Davies GM, Ellwood RP, Kay EJ: Prevention. Part 4: what advice should be given to patients? Br Dent J 2003;195:135–141.

59 Hawkins R, Locker D, Noble J, Kay EJ: Prevention. Part 7: professionally applied topical fluorides for caries prevention. Br Dent J 2003;195:313–317.

Prof. Juliano Pelim Pessan
Department of Pediatric Dentistry and Public Health, Araçatuba Dental School, São Paulo State University
Rua José Bonifácio, 1193
16015–050 Araçatuba – SP (Brazil)
Tel. +55 18 3636 3274, E-Mail jpessan@foa.unesp.br

Buzalaf MAR (ed): Fluoride and the Oral Environment.
Monogr Oral Sci. Basel, Karger, 2011, vol 22, pp 133–145

Systemic Fluoride

Fábio Correia Sampaio[a] · Steven Marc Levy[b]

[a]Department of Clinical and Social Dentistry, Health Science Centre, Federal University of Paraiba, João Pessoa, Brazil; [b]Department of Preventive and Community Dentistry and College of Public Health, Department of Epidemiology, College of Dentistry, University of Iowa, Iowa City, Iowa, USA

Abstract

There is substantial evidence that fluoride, through different applications and formulas, works to control caries development. The first observations of fluoride's effects on dental caries were linked to fluoride naturally present in the drinking water, and then from controlled water fluoridation programs. Other systemic methods to deliver fluoride were later suggested, including dietary fluoride supplements such as salt and milk. These systemic methods are now being questioned due to the fact that many studies have indicated that fluoride's action relies mainly on its post-eruptive effect from topical contact with the tooth structure. It is known that even the methods of delivering fluoride known as 'systemic' act mainly through a topical effect when they are in contact with the teeth. The effectiveness of water fluoridation in many geographic areas is lower than in previous eras due to the widespread use of other fluoride modalities. Nevertheless, this evidence should not be interpreted as an indication that systemic methods are no longer relevant ways to deliver fluoride on an individual basis or for collective health programs. Caution must be taken to avoid excess ingestion of fluoride when prescribing dietary fluoride supplements for children in order to minimize the risk of dental fluorosis, particularly if there are other relevant sources of fluoride intake – such as drinking water, salt or milk and/or dentifrice. Safe and effective doses of fluoride can be achieved when combining topical and systemic methods.

The idea of ingesting fluoride to prevent and control dental caries is not new – in fact the benefits of ingesting fluoride for this purpose have been suggested since the 1870s [1]. At that time, fluoride lozenges were recommended in England, and later in Germany and Scandinavia. In the beginning of the 1900s, the product 'fluoridens' (calcium fluoride, Cross & Co.) was available in several European countries and the USA. The lozenges were probably not very popular among dental practitioners, and they were therefore forgotten [2, 3]. As a result, the strategy of using fluoride by ingestion to control carious lesions was also forgotten until the 1930s, when water fluoridation gained status as a reasonable strategy to reduce the prevalence of dental caries [4].

In the beginning of the 20th century, brown stains (the so-called 'mottled enamel') in subjects from Colorado Springs (USA) intrigued the American dentist Frederick McKay. The observation of similar stains in individuals from other areas of the USA led him to conclude that some substance in the water was responsible for the high prevalence of 'mottled enamel'. After advanced chemical analyses of many samples of drinking water, the substance was identified as fluoride and the term 'mottled enamel' became 'dental fluorosis', thus relating the dental condition to its causative agent [5, 6]. As a result, at that time, fluoride was a chemical compound better known for causing stains in the teeth rather than a substance for controlling dental caries. Moreover, the first investigations exploring the fluoride/dental fluorosis relationship indicated that excessive ingestion of fluoride from the drinking water could lead to dental fluorosis in a dose-response pattern [5, 7].

Meanwhile, a series of epidemiological investigations known as the '21-city study' conducted by Trendley Dean concluded that water with 1 mg/l (which is equivalent to 1 ppm) of natural fluoride could provide protection against dental caries with minimal prevalence of dental fluorosis, and with almost all of it mild (only opacity, not staining). This impressive observation eventually triggered the water fluoridation trials in the USA and officially introduced systemic fluoride methods into dentistry [6, 8, 9].

The first experiences with water fluoridation in Grand Rapids in 1945, and later in Newburgh and Kingston, provided relevant results that clearly indicated a caries reduction of about 60% in children living in the cities with fluoride in the drinking water versus those without fluoridated water [3, 10]. Since the ingestion of excessive fluoride in water could lead to dental fluorosis, then came the question: what would be the effect of a lower ingestion of fluoride? It was suggested that lowering the amount of fluoride ingested would provide caries protection without producing objectionable dental fluorosis. The concept that fluoride's preventive action comes from being ingested and incorporated into tooth mineral became the obvious mechanism of action [3, 6, 11].

All these historical facts are important for understanding the great popularity of systemic methods of delivering fluoride from the 1950s until the end of the 1980s, when understanding of its mechanism of action shifted to the primary topical effect rather than the systemic one [see Buzalaf et al., this vol., pp. 97–114]. In addition to the acceptance of a strong systemic effect, the outstanding success of these first publications on water fluoridation created the belief that the beneficial effects of this method would endure for a lifetime, and this turned out to not be the case [12].

Fluoride remains the cornerstone of modern non-invasive dental caries prevention and management, but its mechanism of action remains a matter of debate [13]. Bibby et al. [14] conducted a study in the 1950s comparing the efficacy of fluoride lozenges intended to be sucked with coated fluoride pills intended to be swallowed. The results of this trial demonstrated that the group using lozenges had fewer carious lesions compared with the group that used the coated pills. This study provided clear evidence that the mechanism of action of fluoride is mostly post-eruptive [11, 13]. Several other studies confirmed the scientific evidence favoring the primacy of fluoride's post-eruptive over pre-eruptive effects in cariostasis [12]. These observations have led to a rethinking of the rationale for using a systemic method of delivering fluoride on an individual or collective basis. Even though systemic fluoride methods were originally designed to promote caries protection by ingestion, anti-caries benefits are delivered primarily through topical effects due to the direct contact of fluoride on the tooth surface prior to ingestion. The beneficial effect can also be explained by the fluoride ingested that returns to the oral cavity when incorporated into the saliva [see Buzalaf et al., this vol., pp. 97–114]. Hence, in order to obtain the maximum benefits of fluoride from systemic methods, continued exposure has to occur. Interruption of systemic fluoride methods can jeopardize the beneficial effects of fluoride in reducing caries incidence [15].

The salts most often used for systemic fluoridation are presented in table 1. The choice and formulas depend on the dose, stability and practicality

Table 1. Fluoride compounds and concentrations usually used in different systemic methods of fluoridation

Fluoridation methods	Fluoride compounds	Fluoride concentrations
Water fluoridation	hydrofluorosilicate (FSA), sodium fluorosilicate, sodium fluoride	0.7–1.2 mg/l
Salt fluoridation	potassium fluoride, sodium fluoride	250–300 mg/kg
Milk fluoridation	sodium fluoride or disodium monofluorophosphate	5 mg/l
Dietary fluoride supplements	sodium fluoride, acidulated phosphate fluoride, potassium fluoride, calcium fluoride	0.25–1.0 mg/day

[16]. The best-known systemic fluoride method is water fluoridation. However, it is also important to recognize that other ways of delivering fluoride – such as salt fluoridation, milk fluoridation and fluoride-containing supplements – still have their importance in dentistry. This is particularly relevant in developing countries where access to oral health care can be very limited.

Water Fluoridation

Recent publications have suggested that over 300 million people in almost 40 countries are exposed to fluoride from adjusted fluoridated water supplies [17, 18]. In the USA alone, an estimated 195 million people (approximately 72.4% of the population) are currently receiving the benefits of optimally fluoridated water [19, 20]. Considering these estimates, more than half of the population of the world receiving the benefits of water fluoridation live in the USA. A national health promotion and disease prevention initiative in the USA, known as 'Healthy People 2010', included an objective to increase the proportion of the US population served by community water systems with optimally fluoridated water to 75% [21]. For 2020, it is expected that nearly 80% of Americans will be served by community water fluoridation [22].

Water fluoridation is a rather simple technique that consists of adding a controlled amount of fluoride to the water supply in concentrations ranging from 0.7 to 1.2 mg/l (equivalent to 0.7–1.2 ppm) depending on the local average temperature. In warm climate regions, the recommended concentrations of fluoride are low (0.7 mg/l) due to higher consumption of water, whereas in more temperate climate regions the concentration can be higher since water consumption is lower [16, 23, 24]. Recently, the Department of Health and Human Services of the United States proposed a new standardized level of 0.7 ppm fluoride throughout the country as an appropriate level for maximizing benefits while minimizing any risks associated with excess ingestion [25].

There is no central source that collects and updates information on costs of water fluoridation programs all over the world. Nevertheless, the average annual cost of this method per person has been estimated to range from USD 0.10 up to USD 5.41, which makes fluoridation a very cost-effective measure for reducing dental caries [16, 26]. Water fluoridation can be regarded as a low-cost method to deliver fluoride, particularly for those communities where oral health care and particularly fluoride dentifrices are not available and/or not affordable. Variables that influence the costs per capita of a fluoridation project include: (1) the size of the community (the smaller the community, the higher the per capita cost); (2) the prevalence of dental caries in the population; (3) the number of water sources; (4) the type of equipment; (5) the fluoride compound; (6) the availability of technical support.

The most common fluoride compounds used are hexafluorosilicic acid (H_2SiF_6), also known as fluorosilicic acid, which comes in a liquid form, and disodium hexafluorosilicate (Na_2SiF_6), also known as sodium fluorosilicate (a powder). It must be emphasized that these compounds are not from industrial waste. Fluorosilicic acid is more frequently used. When it is introduced into the water system, it dissociates to release fluoride ions into the water. This process is similar to what happens to the fluoride ion when it is naturally present in the water supplies. Whitford et al. [27] observed that the major features of human fluoride metabolism are not affected by different chemical compounds commonly used to fluoridate water, or whether the fluoride is present naturally or added artificially. Hence, there is no difference chemically between natural and artificial fluoridation [28].

The US Centers for Disease Control and Prevention recognized fluoridation as one of the major public health measures of the 20th century [29]. Several other organizations – such as the World Health Organization [30] and the American Dental Association [31] – have recognized the effectiveness of water fluoridation in reducing the prevalence of dental caries. In spite of this, water fluoridation is frequently questioned by anti-fluoridationists who cite freedom-of-choice issues or the potential dangers to humans from fluoride [32]. However, there is no evidence of harmful effects of fluoride related to optimal water fluoridation, with the exception of the potential to increase the prevalence of dental fluorosis [24, 29, 33, 34].

Although water fluoridation has been demonstrated to be effective and safe, anti-fluoridationist challenges persist in the USA and elsewhere. Some European countries were not successful in establishing or maintaining water fluoridation programs

due to political or legal reasons [15]. Another important issue related to water fluoridation is the increased consumption of bottled water over the last two decades, even in countries where tap water is considered safe and of excellent quality. The potential causes for this behavior are dissatisfaction with tap water organoleptics (especially taste) and general health risk concerns [35]. Nevertheless, water fluoridation is still an important way to deliver fluoride to control caries [10, 18, 33].

The multiple sources of fluoride and expansion of fluoride therapies have created a complex scenario for evaluating total fluoride ingestion and isolating the beneficial effects of water fluoridation. Hence, the magnitude of benefit from water fluoridation is no longer 50–70%, as found in early studies when caries was prevalent in many parts of the world and water fluoridation was the main source of fluoride [28, 36, 37]. This beneficial effect is certainly lower than in previous years and probably it is also confounded by other methods of fluoride delivery, especially the topical ones [13]. Exposure of individuals to many sources of fluoride has raised concern about the potential to increase the prevalence of dental fluorosis [see Buzalaf and Levy, this vol., pp. 1–19]. It must be pointed out that the majority of the dental fluorosis cases related to water fluoridation are at very mild or mild levels, and are often not even noticed by those who are affected. Thus, the prevalence of aesthetically objectionable dental fluorosis due to water fluoridation is low, and does not represent a health issue [15, 38–40].

Several studies have compared caries prevalence in towns with and without water fluoridation programs or after fluoridation cessation. Some studies reported that caries prevalence remained almost the same after water fluoridation cessation [12, 13, 41]. More aggressive use of sealants and other fluorides may party explain the results in some regions of the world. However, these findings need to be taken with caution, since there is still a noticeable effect of water fluoridation in reducing carious lesions incidence in other parts of the world [15]. Recent data from the National Brazilian Epidemiologic Survey showed a 30–40% lower decayed (D) component of the DMFT in 12-year-old children from fluoridated towns where water fluoridation had been implemented for more than 5 years [42]. It is interesting to note that high-concentration fluoride dentifrices (1,500 ppm) have been used widely since the 1990s in both fluoridated and non-fluoridated towns in the country. In preliminary analyses of 2010 data from several communities in Brazil, most of them capitals of their states, a water fluoridation preventive effect of about 20–30% is still present when comparing fluoridated with non-fluoridated towns (raw data from the National Brazilian Epidemiologic Survey, 2010). These results are similar to those observed in the USA in the mid-1980s [43], and in line with data obtained recently in Australia [39]. In this later study, a 3-year follow-up of caries status was carried out in adolescents. The effectiveness of water fluoridation was observed even in the presence of the effect of fluoride from other methods.

Several studies have shown that water fluoridation reduced the prevalence of caries and the number of affected teeth, as well as social inequalities, among groups with a different socioeconomic status [44–50]. However, the features of social complexity of the populations in each country complicate these comparisons. For instance, in New Zealand and Finland, water fluoridation had a similar beneficial effect for all social classes, whereas in some other countries this was not the case [51–53]. In a Brazilian study carried out in the wealthiest part of the country, communities with better social status had lower caries experience, probably due to the mediating effect of receiving fluoridation earlier than in other areas [49]. In general, there is a consensus that water fluoridation can be most advantageous for more deprived communities where other health policies are less available [10, 54]. On the other hand, one must bear in mind that in underprivileged communities there are limitations even in

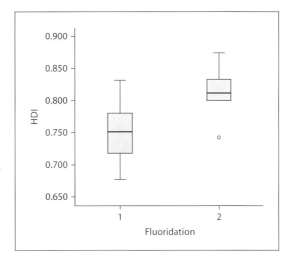

Fig. 1. Human development index (HDI) in Brazilian States where water fluoridation does not reach 40% of the territory (1) and in those where water fluoridation is available in more than 60% of the cities (2).

the access to potable water. These regions generally have certain similarities: technical capabilities are limited and political support for water fluoridation is often less favorable.

Human development and potable water availability has been subject of study of many investigations [55]. Evaluation of the impact of water accessibility can provide relevant data on water quality and health parameters as well. The Human Development Index (HDI) has been successfully used as suitable ranking of development of countries and standard of living in target regions within countries. This index is a composite statistic reference number (from zero = lowest, to one = highest level) which is calculated based on life expectancy, education and standard of living [56]. The relationship of water fluoridation to social inequalities can be evaluated by the HDI. This is particularly important for developing countries where water fluoridation is feasible. Figure 1 shows the HDI in Brazilian States where water fluoridation does not reach 40% of the territory and in those where water fluoridation is available in more than 60% of the cities. There is a clear indication that water fluoridation reaches the most privileged groups (HDI >0.80) more efficiently than it does those who are in greater need of this health benefit (HDI <0.75). Since potable water is a basic unmet need in many underprivileged populations in developing countries, one potential strategy for expanding water fluoridation in such areas might be linking fluoridation projects with the need for potable water – though implementation of this plan may face limitations in certain geographical areas.

In summary, water fluoridation remains the most cost-effective way to deliver fluoride for prevention and control of caries at the community level. Opponents of community water fluoridation have been overstating adverse health effects, including concerns with aesthetically objectable dental fluorosis, without scientific basis. Community water fluoridation can be the most important and sometimes the only feasible oral health program for some underprivileged groups. Although the currently measured percentage level of effectiveness of water fluoridation in many areas is lower than in previous eras due to the more widespread use of other fluoride modalities, its importance as a general health measure must not be underestimated.

Salt Fluoridation

Salt fluoridation (at a concentration of about 250–350 mg/kg) can be considered as an alternative to fluoridation of drinking water. It was introduced in Switzerland in the 1950s based on the success of the use of iodized salt to prevent goiter [3, 57, 58]. As mentioned in the introduction to this chapter, the objective of any fluoridation method in the 1950s was to promote the ingestion of fluoride in order to achieve its cariostatic effect. Hence, the concept of using salt fluoridation in a community has a different aim today, which is to reach communities and regions in the world where oral care

prevention measures, and particularly fluoride toothpastes, are not available [59].

The first epidemiological studies to evaluate the effectiveness of fluoridated salt in reducing caries prevalence were performed in Colombia, Hungary and Switzerland [58]. The outcomes of these studies indicated that salt fluoridation generally showed very similar beneficial results to those observed for water fluoridation [60].

Recent statistics indicate that salt fluoridation is available in nearly all Latin American countries, except Brazil, Chile and Panama. It is still available in several European countries, including France, Germany and Switzerland. There are national regulations or authorizations for the production and marketing of fluoridated salt in eight European countries: Austria, Czech Republic, France, Germany, Romania, Slovakia, Spain and Switzerland [58, 59, 61–63].

The costs for implementing salt fluoridation are similar to those for water fluoridation regarding the equipment for initial operation. However, during operation, salt fluoridation has an estimated cost 10–100 times lower than that associated with water fluoridation programs. According to Gillespie and Marthaler [64], the costs of salt fluoridation can vary from USD 0.015 up to USD 0.030 per capita/year, which is so low that many producers do not raise the price of the product after fluoridation is implemented.

In contrast with water fluoridation, which is readily available to the whole community, salt fluoridation can provide a choice for the consumer. According to Jones et al. [65], the individual choice is one positive aspect of a fluoridated salt program, since it can be sold alongside a non-fluoridated alternative. Individual choice makes salt fluoridation more acceptable for some people from ethical and social policy perspectives. On the other hand, it can weaken its caries-preventive impact since salt is not used similarly on an individual basis [16]. Another aspect to consider is that many variants of the commercial distribution or 'channels' to reach the consumer may exist. These channels include: domestic salt, meals at schools, large kitchens, and in food items such as bread. In Switzerland and other Latin American countries, all commercial channels are utilized. In some other European countries (France and Germany), the salt fluoridation program is mainly based on domestic salt [62, 65].

One point of concern is that promoting salt fluoridation could be contraindicated from the perspective of general public health, since greater salt consumption is linked to hypertension. However, according to Jones et al. [65] people do not need to change their usual behavior to benefit, and if a secular decline in salt consumption were to take place, an increase in fluoride concentration could be considered. To support this view, Bürgi and Zimmerman [57] expressed the opinion that preventing hypertension through restricting salt intake and eliminating iodine deficiency through iodized salt are not in conflict. It is estimated that among communities or groups usually consuming low-salt diets (<5 g NaCl per person per day), essential hypertension will be uncommon. Moreover, there is no doubt that some salt is required by man, and estimates of normal daily requirements for adults have ranged up to 15 g per day [66].

The fluoride compounds used are usually sodium fluoride and potassium fluoride (table 1), which are included in the salt during manufacture of the product. A wide range of concentrations of fluoride have been tested with concentrations varying from 90 mg/kg up to 350 mg/kg. Most programs have used 250 or 350 mg/kg, and some studies have suggested that the ideal concentration of fluoride in the salt should be about 250 mg/kg [58, 60]. This level of fluoride was supported by a study that measured salivary fluoride after a meal prepared with fluoridated salt at 250 mg/kg. The results of this trial indicated that, at this concentration, the level of fluoride in saliva was very similar to that found in the saliva of individuals exposed to water fluoridation at 1 mg/l [67]. Most of the studies designed for monitoring salt

fluoridation use urine as a biomarker [see Rugg-Gunn et al., this vol., pp. 37–51]. There are several studies that show that ingestion of fluoridated salt can increase fluoride excretion, and consequently this can be a useful way to monitor compliance of individuals with a salt fluoridation program, as well as a good alternative to monitor possible excessive fluoride ingestion [58, 68].

One interesting experience using salt fluoridation was noted in Jamaica, where a salt fluoridation program started in 1987. At that time, the mean DMFT of 12-year-old children was 6.7 (very severe) and recently the DMFT was about 1.1 (low) [69]. The salt fluoridation program was considered appropriate for the island due to geographical conditions, the low concentrations of water-borne fluoride (which do not exceed 0.3 mg/l), and the availability of bottled water also having the same levels of fluoride. A recent study observed that 96% of rural and 100% of urban Jamaican children in the sample were consuming fluoridated salt [59].

Similar to issues raised with regard to fluoridated water, there has been some concern about the simultaneous combination of fluoride ingested from both dentifrice and salt. Available data suggest that this combination has not resulted in objectionable enamel fluorosis levels [70, 71]. However, increased mild dental fluorosis was observed in children who used fluoride tablets in association with fluoridated salt [37].

Although salt fluoridation has received support from official health agencies such as the World Health Organization, regular ongoing surveillance of fluoride concentrations in the salt is necessary [59]. The concentration of 250–300 mg/kg of fluoride in salt is regarded as the ideal concentration, while the concentration of 200 mg/kg of fluoride is regarded as the minimal acceptable level of fluoride in salt to achieve a meaningful effect on caries control. Salt fluoridation should be considered when water fluoridation is technically difficult or due to economic or sociocultural reasons it cannot be implemented. In summary, the advantages of using salt as a vehicle for delivering fluoride outweigh the drawbacks related to this method, such as variation in ingestion, difficulties in maintaining the ideal concentration and concerns with hypertension.

Milk Fluoridation

The idea of introducing fluoride into milk was first published in 1953 [72], and the first outcomes of a clinical trial with milk fluoridation were available in 1959 [73, 74]. Milk fluoridation became more popular only decades later when a charitable foundation started to promote this method [75]. The first community milk fluoridation program was implemented in Bulgaria in the cities of Plovdiv and Asenovgrad in 1988, reaching 15,000 children [74]. In the 1990s, milk fluoridation projects were implemented in Russia, China, Chile, Peru and the UK. However, Peru later ceased milk fluoridation due to the introduction of salt fluoridation. In 2000, a milk fluoridation program was introduced in Thailand, and recently the Republic of Macedonia started milk fluoridation [75].

It is estimated that about 800,000 children are receiving fluoridated milk [75]. Most data available for this method are from studies with children, since the child population is the target age group with school-based programs. Milk consumption varies considerably when comparing different regions of the world. For instance, the worldwide average consumption of milk was estimated to be 78 kg per person/year, but consumption was highest in industrialized countries (212 kg per person/year) when compared with developing countries (45 kg per person/year) [74]. Latin America has the highest estimates among developing countries with 110 kg per person/year. There has been a modest increase in milk consumption in most industrialized countries over the last 30 years, but in Western European countries a decline was observed during the same period. In 2006, Finland had the highest consumption per capita of milk,

with an impressive average of 183.9 liters, followed by Sweden with 145.5 liters. On the other hand, in China only 8.8 liters per capita were consumed [76]. It is predicted that the worldwide average consumption of milk could reach 90 kg per person/year in the year of 2030 [74].

Sodium fluoride or disodium monofluorophosphate are the fluoridating compounds included in milk (table 1). The manufacture of fluoridated milk involves simple production techniques regardless of its various forms (pasteurized, sterilized, UHT or powdered) [74, 77]. All the products have been shown to be stable, with considerable amounts of fluoride remaining throughout their shelf-life. The rationale for ingestion of fluoridated milk is that it increases the concentration of fluoride in saliva to levels similar to those observed for optimally fluoridated water. Considering the amounts of water and milk usually consumed, in terms of caries prevention the fluoride concentration equivalent to 1 mg/l of fluoride in water is 5 mg/l of fluoride in milk [74, 78].

The main constituents of whole cow's milk, other than carbohydrate (4.5%), considered relevant for the de- and remineralization processes are fat (up to 3.9%), protein (3%), phosphorus (92 mg/100 g) and calcium (118 mg/100 g). The amount of carbohydrate present in milk would be sufficient to classify this food item as cariogenic. Nevertheless, lactose is regarded as the least cariogenic of the common dietary sugars [77]. Studies focusing on the positive aspects of having milk as a vehicle for fluoride delivery have indicated that milk would appear to reduce the cariogenic potential of dental plaque due to: (1) lactose being the least cariogenic of dietary sugars; (2) the protective role of casein, and possibly fats; (3) the protective role of calcium and phosphorus in the de- and remineralization processes. However, the favorable features of milk can be strongly compromised when sucrose is added [79]. In fact, cow's milk is essentially non-cariogenic, but the addition of sucrose in the milk can promote early caries in young children [80].

A systematic review on fluoridated milk concluded that high-quality randomized clinical trials concerning the effectiveness of fluoridated milk for caries prevention are lacking. This gap compromises the evidence for supporting fluoridated milk as an effective method for caries control [81]. However, from the few available studies, it can be concluded that children should begin to drink fluoridated milk at an early age, preferably before 4 years, in order to reduce caries in their primary teeth. In addition, children should be drinking fluoridated milk when their first permanent molars erupt in order to help protect these teeth.

Dietary Fluoride Supplements

The use of dietary fluoride supplements for controlling dental caries was introduced by the end of the 1940s when it was assumed that the 'systemic' effect of fluoride was its main mechanism of action. It is important to note that the dietary fluoride supplements were introduced before the widespread use of fluoride dentifrices, varnishes, gels and other professional methods for applying fluoride. Thus, when dietary fluoride supplements were introduced, the major source of fluoride was the drinking water [16]. Since that time, dietary fluoride supplements have been intended to substitute for fluoridated water in areas where water fluoridation is not available or feasible.

In the beginning of the 1950s, the American Dental Association recommended supplementing the domestic drinking water with of 0.25 mg/l fluoride for children up to 6 years of age. The recommendation was to mix fluoride with the water in a bottle that would be stored in the refrigerator. The children were asked to drink this fluoridated water and eat food prepared with it [82].

The term dietary fluoride supplements can mean different forms of manufactured products: (1) tablets or drops (to be swallowed); (2) tablets for chewing; (3) lozenges (to be sucked or dissolved in the mouth) [16]. The different fluoride compounds used as supplements are shown in table 1. The most common type of fluoride used is sodium fluoride.

Several studies support the view that the major beneficial cariostatic effect of fluoride supplements is due to its post-eruptive effect [12, 81, 83]. The pre-eruptive effect of dietary fluoride supplements has been questioned and received little credit. Moreover, the pre-eruptive effect and the early use of supplements have been linked to dental fluorosis in children. As a result, in some countries (e.g. Canada), dietary fluoride supplements are not indicated for most citizens and particularly for those before the third year of age, because of the potential risk for dental fluorosis [16, 83].

The balance between the beneficial effect on caries prevention and the potential increased prevalence of dental fluorosis is the key point in considering the use of fluoride supplements. This balance between benefits and risks can be influenced by the child's age and caries risk status. It is also important to point out the fluoride concentration of the child's primary sources of drinking water as an important variable [84].

In spite of the potential risk for dental fluorosis, dietary fluoride supplements are regarded as effective in preventing caries and are still available in several countries [85]. In addition, the current evidence indicates that the incidence of carious lesions can be reduced in both the primary and permanent dentition by the regular use of dietary fluoride supplements [86]. However, a recent systematic review indicated that the effectiveness of fluoride supplements is weak for the primary dentition [83]. Due to the fact that dietary fluoride supplements can increase children's intake of fluoride, this method is recommended only for groups of or individual children at high caries risk. These recommendations emphasize that, for an appropriate prescription of dietary fluoride supplements, there is a need for caries risk assessment in association with some estimate of the total fluoride intake. In summary, the balance between the caries-preventive benefits of dietary

Table 2. Fluoride supplement dosage schedule for 2010 (mg/day)

	Fluoride ion level in drinking water		
	<0.3 mg/l	0.3–0.6 mg/l	>0.6 mg/l
Age			
Birth to 6 months	none	none	none
6 months to 3 years	0.25	none	none
3–6 years	0.50	0.25	none
6–16 years	1.0 mg	0.50	none

Approved by the American Dental Association, American Academy of Pediatrics and American Academy of Pediatric Dentistry.

fluoride supplementation and the risk of dental fluorosis has to be evaluated for the appropriate implementation of this method at both group and individual levels.

In addition to child's age and caries risk status, the sources of fluoride intake such as drinking water have to be considered when prescribing dietary fluoride supplements. Hence, it is recommended that the fluoride content in the water should be determined, whether a public water source or bottled water is the primary source of drinking water. There is evidence that the use of both fluoridated salt and fluoride tablets is more effective in reducing caries in children when compared with the use of fluoridated salt alone [37]. However, similarly to what happens in water fluoridation, the fluoride intake from salt and tablets simultaneously can increase the occurrence of mild fluorosis in permanent incisors [37].

Since 1994, the American Dental Association has recommended that fluoride supplements should not be given from birth up to 6 months (table 2) [87]. This is an interesting precaution for avoiding significantly increased fluoride intake during the early stages of tooth formation, and thus reducing the risk of dental fluorosis. Finally, there is no doubt that dietary fluoride supplements can play a role in caries control, particularly for children at elevated risk of caries and

not receiving other systemic or topical fluoride modalities. Because dietary fluoride supplements seem to be more effective in preventing caries in the permanent dentition, and that their use before the age of 6 years (but especially before the age of 3 years) is associated with an increased prevalence of dental fluorosis, they should be prescribed only for high-caries-risk groups and individuals from the age of 3 years on. In addition, they can be well suited to some remote populations not receiving other methods of fluoride delivery and in areas where this method is accepted and compliance is achieved [88].

Conclusion

Concerning the systemic effect of fluoride, most studies support the view that the caries-preventive effect of fluoride is mainly post-eruptive. This evidence should not be interpreted as a limitation of systemic fluoride methods, since some topical effect cannot be disregarded when someone is ingesting fluoride in water, milk or salt or sucking a fluoride lozenge. However, since multiple sources of fluoride exposure and ingestion may exist, a well coordinated approach for promoting fluoride delivery is essential, particularly if the individual can be exposed to different systemic and topical methods simultaneously. In fact, the classification of fluoride delivery methods into 'systemic' and 'topical' is misleading, since it is recognized that the topical effect is more important. A more rational classification would be as follows: individual methods (fluoride dentifrices, mouth rinses and supplements), collective methods (fluoridated water, salt and milk) and professional methods (fluoride varnishes, gels, solutions and foams). This approach can simplify the consideration of benefits and risks when combining different fluoride methods, particularly individual and collective ones.

There is strong evidence that fluoride under different applications and formulas works for controlling caries development. Nevertheless, there is a lack of high-quality clinical trials concerning the caries-preventive effect of fluoridated milk and salt.

Regarding dietary fluoride supplements, there is strong evidence for an increased risk of dental fluorosis when they are used before 3 years of age. Moreover, evaluation of additional sources of fluoride is important when using fluoride supplements.

To date, there is strong evidence that water fluoridation is still an effective method of caries prevention. However, technical requirements for implementing water fluoridation may pose limitations for some developing countries, resulting in water fluoridation not being available where it is most needed.

Finally, it must be pointed out that fluoride has been used to prevent caries for more than 60 years, and several clinical and laboratory studies have demonstrated that fluoride used in both topical and systemic ways has the ability to control caries. Thus, fluoride is still of the utmost importance for preventing dental caries, as well as for modern non-invasive dental caries management.

References

1 Hunstadbraten K: Fluoride in caries prophylaxis at the turn of the century. Bull Hist Dent 1982;30:117–120.

2 Hoffmann-Axthelm W: History of Dentistry. Chicago, Quintessence, 1981, pp 318–322.

3 Murray JJ, Rugg-Gunn AJ, Jenkins GN: Fluorides in Caries Prevention, ed 3. Oxford, Butterworth-Heinemann, 1991.

4 Maier FJ: Manual of Water Fluoridation Practice. New York, McGraw-Hill, 1963.

5 Dean HT: Endemic fluorosis and its relation to dental caries. Public Health Rep 1938;53:1443–1452.

6 McClure FJ: Water Fluoridation: The Search and the Victory. Bethesda, National Institutes of Health, 1970.

7 McKay FS: The relation of mottled enamel to caries. J Am Dent Assoc 1933;20:1137–1149.

8 Horowitz HS: Effectiveness of School Water Fluoridation and Dietary Fluoride Supplements in School-aged Children J Publ Hlth Dent 1989;49(5 Spec No): 290–296.

9 Horowitz HS: The effectiveness of community water fluoridation in the United States. J Public Health Dent 1996;56:253–258.

10 Centers for Disease Control and Prevention: Achievements in Public Health,1990–1999: fluoridation of drinking water to prevent dental caries. Morb Mortal Wkly Rep 1999;48:933.

11 Fejerskov O: Changing paradigms in concepts on dental caries: consequences for oral health care. Caries Res 2004;38:182–191.

12 Limeback H: A re-examination of the pre-eruptive and post-eruptive mechanism of the anti-caries effects of fluoride: is there any anti-caries benefit from swallowing fluoride? Community Dent Oral Epidemiol. 1999;27:62–71.

13 Hellwig E, Lennon AM: Systemic versus topical fluoride. Caries Res 2004;38: 258–262.

14 Bibby BG, Wilkins E, Witol E: A preliminary study of the effects of fluoride lozenges and pills on dental caries. Oral Surg Oral Med Oral Pathol. 1955;8:213–216.

15 Newbrun E: What we know and do not know about fluoride. J Public Health Dent 2010;70:227–233.

16 Fejerskov O, Ekstrand J, Burt B: Fluoride in Dentistry, ed 2. Copenhagen, Munksgaard, 1996.

17 Browne D, Whelton H, O'Mullane D: Fluoride metabolism and fluorosis. J Dent 2005;33:177–186.

18 Pizzo G, Piscopo M, Pizzo I, Giuliana G: Community water fluoridation and caries prevention: a critical review. Clin Oral Invest 2007;11:189–193.

19 Bailey W, Barker L, Duchon K, Maas W: Populations receiving optimally fluoridated public drinking water – United States, 1992–1996. MMWR 2009;57:737–741.

20 Center for Disease Control and Prevention: 2008 Water Fluoridation Statistics. Atlanta, CDC, 2010. www.cdc.gov/fluoridation/statistics/2008stats.htm.

21 Center for Disease Control and Prevention: About Healthy People. Atlanta, CDC, 2010. www.cdc.gov/nchs/healthy_people/hp2010.htm.

22 Healthy People: Oral Health. www.healthypeople.gov/2020/topicsobjectives2020/overview.aspx?topicid=32.

23 Murray JJ: Efficacy of preventive agents for dental caries. Systemic fluorides: water fluoridation. Caries Res 1993; 27(suppl 1):2–8.

24 Whitford GM: The Metabolism and Toxicology of Fluoride, ed 2. Basel, Karger, 1996.

25 Department of Health and Human Services: Proposed HHS recommendation for fluoride concentration in drinking water for prevention of dental caries. Federal Register Notices 2011;76: 2383–2388.

26 Ciketic S, Hayatbakhsh MR, Doran CM: Drinking water fluoridation in South East Queensland: a cost-effectiveness evaluation. Health Promot J Austr 2010;21:51–56.

27 Whitford GM, Sampaio FC, Pinto CS, Maria AG, Cardoso VES, Buzalaf MAR: Pharmacokinetics of ingested fluoride: lack of effect of chemical compound. Arch Oral Biol 2008;53:1037–1041.

28 Lennon MA: One in a million: the first community trial of water fluoridation. Bull World Health Organ 2006;84: 759–760.

29 Center for Disease Control and Prevention: Recommendations for using fluoride to prevent and control dental caries in the United States. MMWR Morb Mortal Wkly Rep 2001;50:1–42.

30 Fawell J, Bailey K, Chilton J, Dahi E, Fewtrell L, Magara Y: Fluoride in drinking-water. London, IWA (World Health Organization publication), 2006

31 American Dental Association: Fluoride & Fluoridation. Chicago, ADA, 2010. www.ada.org/fluoride.aspx.

32 Cross DW, Carton RJ: Fluoridation: a violation of medical ethics and human rights. Int J Occup Environ Health 2003;9:24–29.

33 McDonagh MS, Whiting PF, Wilson PM, Sutton AJ, Chestnutt I, Cooper J, Misso K, Bradley M, Treasure E, Kleijnen J: Systematic review of water fluoridation. BMJ 2000;321:855–859.

34 Committee on Fluoride in Drinking Water, Board on Environmental Studies and Toxicology, Division on Earth and Life Studies, National Research Council of the National Academies: Fluoride in Drinking Water: A Scientific Review of EPA's Standards. Washington, National Academies Press, 2006.

35 Doria MF: Bottled water versus tap water: understanding consumers' preferences. J Water Health 2006;4:271–276.

36 Levy SM: An update on fluorides and fluorosis. J Can Dent Assoc 2003;69: 286–291.

37 Meyer-Lueckel H, Grundmann E, Stang A: Effects of fluoride tablets on caries and fluorosis occurrence among 6- to 9-year olds using fluoridated salt. Community Dent Oral Epidemiol 2010;38: 315–323.

38 Griffin SO, Beltrán ED, Lockwood SA, Barker LK: Esthetically objectionable fluorosis attributable to water fluoridation. Community Dent Oral Epidemiol 2002;30:199–209.

39 Spencer AJ, Armfield JM, Slade GD: Exposure to water fluoridation and caries increment. Community Dent Health 2008;25:12–22.

40 Chankanka O, Levy SM, Warren JJ, Chalmers JM: A literature review of aesthetic perceptions of dental fluorosis and relationships with psychosocial aspects/oral health-related quality of life. Community Dent Oral Epidemiol 2010;38: 97–109.

41 Maupomé G, Shulman JD, Clark DC, Levy SM, Berkowitz J: Tooth-surface progression and reversal changes in fluoridated and no-longer-fluoridated communities over a 3-year period. Caries Res 2001;35:95–105.

42 Ministério da Saúde: Projecto SB Brasil 2003: Condições de Saúde Bucal da População Brasileira 2002–2003. Resultados principais (in Portuguese). Brasília, Ministério da Saúde, 2004.

43 Brunelle JA, Carlos JP: Recent trends in dental caries in U.S. children and the effect of water fluoridation. J Dent Res 1990;(69 Spec No):723–727, discussion 820–823.

44 Ellwood RP, O'Mullane DM: The association between area deprivation and dental caries in groups with and without fluoride in their drinking water. Community Dent Health 1995;12:18–22.

45 Lawrence HP, Sheiham A: Caries progression in 12 to 16-year-old schoolchildren in fluoridated and fluoride-deficient areas in Brazil. Community Dent Oral Epidemiol 1997;25:402–411.

46 Riley JC, Lennon MA, Ellwood RP: The effect of water fluoridation and social inequalities on dental caries in 5-year-old children. Int J Epidemiol 1999;28:300–305

47 Jones CM, Worthington H: Water fluoridation, poverty and tooth decay in 12-year-old children. J Dent 2000;28: 389–393.

48 Burt BA: Fluoridation and social equity. J Public Health Dent 2002;62:195–200.

49 Gabardo MC, da Silva WJ, Moysés ST, Moysés SJ: Water fluoridation as a marker for sociodental inequalities. Community Dent Oral Epidemiol 2008; 36:103–107.

50 Fischer TK, Peres KG, Kupek E, Peres MA: Primary dental care indicators: association with socioeconomic status, dental care, water fluoridation and Family Health Program in Southern Brazil (in Portuguese). Rev Bras Epidemiol 2010;13:126–138.

51 Hausen H, Milen A, Heinonen OP, Paunio I: Caries in primary dentition and social class in high and low fluoride areas. Community Dent Oral Epidemiol 1982;10:33–36.

52 Evans RW, Beck DJ, Brown RH, Silva PA: Relationship between fluoridation and socioeconomic status on dental caries experience in 5-year-old New Zealand children. Community Dent Oral Epidemiol 1984;12:5–9.

53 Spencer AJ, Slade GD, Davies M: Water fluoridation in Australia. Community Dent Health 1996;13(suppl 2):27–37.

54 Petersen PE, Bourgeois D, Ogawa H, Estupinan-Day S, Ndiaye C: The global burden of oral diseases and risks to oral health. Bull World Health Organ 2005; 83:661–669.

55 Human Development Report Team, Klugman J, et al: Human Development Report 2010: 20th Anniversary Edition. The Real Wealth of Nations: Pathways to Human Development. New York, Palgrave Macmillan (for United Nations Development Programme), 2010.

56 Webb BC, Simpson SL, Hairston KG: From politics to parity: using a health disparities index to guide legislative efforts for health equity. Am J Public Health 2011;101:554-560.

57 Bürgi H, Zimmermann MB: Salt as a carrier of iodine in iodine deficient areas. Schweiz Monatsschr Zahnmed 2005;115:648–650.

58 Marthaler TM: Overview of salt fluoridation in Switzerland since 1955, a short history. Schweiz Monatsschr Zahnmed 2005;115:651–655.

59 Baez RJ, Marthaler TM, Baez MX, Warpeha RA: Urinary fluoride levels in Jamaican children in 2008, after 21 years of salt fluoridation. Schweiz Monatsschr Zahnmed 2010;120:21–28.

60 Marthaler TM: Increasing the public health effectiveness of fluoridated salt. Schweiz Monatsschr Zahnmed 2005;115:785–792.

61 Tramini P: Salt fluoridation in France since 1986.Schweiz Monatsschr Zahnmed 2005;115:656–658.

62 Schulte AG: Salt fluoridation in Germany since 1991. Schweiz Monatsschr Zahnmed 2005;115:659–662.

63 Gillespie GM, Baez R: Development of salt fluoridation in the Americas. Schweiz Monatsschr Zahnmed 2005;115:663–669.

64 Gillespie GM, Marthaler TM: Cost aspects of salt fluoridation. Schweiz Monatsschr Zahnmed 2005;115:778–784.

65 Jones S, Burt BA, Petersen PE, Lennon MA: The effective use of fluorides in public health. Bull World Health Organ 2005;83:670–676.

66 Dahl LK: Possible role of salt intake in the development of essential hypertension. Int J Epidemiol 2005;34:967–972.

67 Hedman J, Sjöman R, Sjöström I, Twetman S: Fluoride concentration in saliva after consumption of a dinner meal prepared with fluoridated salt. Caries Res 2006;40:158–162.

68 Marthaler TM, Schulte AG: Monitoring salt fluoridation programs through urinary excretion studies. Schweiz Monatsschr Zahnmed 2005;115: 679–684.

69 Baez R, Horowitz H, Warpeha R, Sutherland B, Thamer M: Salt fluoridation and dental caries in Jamaica. Community Dent Oral Epidemiol 2001;29:247–252.

70 Solórzano I, Salas MT, Chavarría P, Beltrán-Aguilar E, Horowitz H: Prevalence and severity of dental caries in Costa Rican schoolchildren: results of the 1999 national survey. Int Dent J 2005;55:24–30.

71 Menghini G: Dental fluorosis in salt fluoridation schemes. Schweiz Monatsschr Zahnmed 2005;115:1026–1030.

72 Ziegler E: Cariesprophylaxe durch Fluorierung der Milch. Schweiz Med Wochenschr 1953;83:723–724.

73 Imamura Y: Treatment of school meals with sodium fluoride as a means of preventing tooth decay. J Oral Dis Acad 1959;26:180–199.

74 Bánóczy J, Petersen PE, Rugg-Gunn AJ: Milk Fluoridation for the Prevention of Dental Caries. Geneva, World Health Organization, 2009.

75 World Health Organization: Oral health promotion: an essential element of a health-promoting school (WHO information series on school health, document 11). Geneva, World Health Organization, 2003.

76 International Dairy Federation: The World Dairy Situation 2007 (423/2007). Brussels, Bulletin of the International Dairy Federation, 2007.

77 Rugg-Gunn AJ: Nutrition and Dental Health. Oxford, Oxford University Press, 1993.

78 Twetman S, Nederfors, Petersson LG: Fluoride concentration in whole saliva and separate gland secretions in schoolchildren after intake of fluoridated milk. Caries Res 1998;32:412–416.

79 Prabhakar AR, Kurthukoti AJ, Gupta P: Cariogenicity and acidogenicity of human milk, plain and sweetened bovine milk: an in vitro study. J Clin Pediatr Dent 2010;34:239–247.

80 Bowen WH, Lawrence RA: Comparison of the cariogenicity of cola, honey, cow milk, human milk, and sucrose. Pediatrics 2005;116:921–926.

81 Espelid I: Caries preventive effect of fluoride in milk, salt and tablets: a literature review. Eur Arch Paediatr Dent 2009;10:149–156.

82 Horowitz HS: Water fluoridation and other methods for delivering systemic fluorides; in Stallard RE (eds): A Textbook of Preventive Dentistry, ed 2. Philadelphia, W.B. Saunders Company, 1982, pp 147–169.

83 Ismail AI, Hasson H: Fluoride supplements, dental caries and fluorosis: a systematic review. J Am Dent Assoc 2008; 139:1457–1468.

84 Levy SM, Broffitt B, Marshall TA, Eichenberger-Gilmore JM, Warren JJ: Associations between fluorosis of permanent incisors and fluoride intake from infant formula, other dietary sources and dentifrice during early childhood. J Am Dent Assoc 2010; 141:1190–1201.

85 Pendrys DG, Haugejorden O, Bårdsen A, Wang NJ, Gustavsen F: The risk of enamel fluorosis and caries among Norwegian children: implications for Norway and the United States. J Am Dent Assoc 2010;141:401–414.

86 Rozier RG, Adair S, Graham F, Iafolla T, Kingman A, Kohn W, Krol D, Levy S, Pollick H, Whitford G, Strock S, Frantsve-Hawley J, Aravamudhan K, Said M, Whall C, Meyer DM: Evidence-based clinical recommendations on the prescription of dietary fluoride supplements for caries prevention: a report of the American Dental Association Council on Scientific Affairs. JADA 2010;141: 1480–1489.

87 American Dental Association: Fluoride supplement dosage schedule – 1994. www.ada.org/3088.aspx?currentTab=1 (accessed 26 December 2010).

88 Gagnon F, Catellier P, Arteau-Gauthier I, Simard-Tremblay E, Lepage-Saucier M, Paradis-Robert N, Michel J, Lavallière A: Compliance with fluoride supplements provided by a dental hygienist in homes of low-income parents of preschool children in Quebec. J Public Health Dent 2007;67:60–63.

Fábio Correia Sampaio
Department of Clinical and Social Dentistry, Health Science Centre
Federal University of Paraiba
João Pessoa, 58051–900 (Brazil)
Tel. +55 83 3216 7795, E-Mail fabios@ccs.ufpb.br

Buzalaf MAR (ed): Fluoride and the Oral Environment.
Monogr Oral Sci. Basel, Karger, 2011, vol 22, pp 146–157

Oral Fluoride Reservoirs and the Prevention of Dental Caries

Gerald Lee Vogel

American Dental Association Foundation, Paffenbarger Research Center, Gaithersburg, Md., USA

Abstract

Current models for increasing the anti-caries effects of fluoride (F) agents emphasize the importance of maintaining a cariostatic concentration of F in oral fluids. The concentration of F in oral fluids is maintained by the release of this ion from bioavailable reservoirs on the teeth, oral mucosa and – most importantly, because of its association with the caries process – dental plaque. Oral F reservoirs appear to be of two types: (1) mineral reservoirs, in particular calcium fluoride or phosphate-contaminated 'calcium-fluoride-like' deposits; (2) biological reservoirs, in particular (with regard to dental plaque) F held to bacteria or bacterial fragments via calcium-fluoride bonds. The fact that all these reservoirs are mediated by calcium implies that their formation is limited by the low concentration of calcium in oral fluids. By using novel procedures which overcome this limitation, the formation of these F reservoirs after topical F application can be greatly increased. Although these increases are associated with substantive increases in salivary and plaque fluid F, and hence a potential increase in cariostatic effect, it is unclear if such changes are related to the increases in the amount of these reservoirs, or changes in the types of F deposits formed. New techniques have been developed for identifying and quantifying these deposits which should prove useful in developing agents that enhance formation of oral F reservoirs with optimum F release characteristics. Such research offers the prospect of decreasing the F content of topical agents while simultaneously increasing their cariostatic effect.

Recent in vitro and in vivo studies as well as clinical observations [1, 2] have emphasized the importance of maintaining a concentration of fluoride (F) in oral fluids that is cariostatically effective. This focus on oral fluid F has in turn drawn attention to the importance of the 'bioavailable' F reservoirs that can persistently increase these concentrations in the fluids surrounding the site of de- and remineralization on the teeth.

Types of Oral Fluoride Reservoirs

Oral F reservoirs can be divided into two broad types, both of which involve calcium (Ca): (1) the mineral deposits of F which include calcium fluoride (CaF_2) and fluorapatite (FAp); (2) biologically/bacterially bound calcium-fluoride (Ca-F) deposits.

Fluorapatite

The formation and dissolution of FAp, Ca_{10} $(PO_4)_6F_2$, is governed by the relationship of a constant, the solubility product (KSP) of this mineral and the ion activity product (IAP) of the ions of which it is composed. The KSP of FAp (KSP_{FAp}) is 10^{-121} [3] while the IAP of FAp can be calculated from $IAP_{FAp} = \{Ca^{2+}\}^{10} \times \{PO_4^{3-}\}^6 \times \{F^-\}^2$ where { } is the chemical activities of these ions. The activity of an ion is the free unbound concentration of the ion multiplied by an ionic-strength-dependent activity coefficient [4]. The relationship between KSP and IAP is a step function: an all-or-nothing relationship in which dissolution/formation of a mineral occurs when the IAP_{FAp} is above the KSP_{FAp} (supersaturation) or below the KSP_{FAp} (undersaturation). Even under cariogenic conditions, IAP_{FAp} values for oral fluid are usually much greater than KSP_{FAp} [5, 6]; thus, FAp does not dissolve, and hence FAp mineral is a poor source of oral fluid F. Furthermore, it is possible, as noted below, to reduce the release of F to oral fluids by inducing the formation of this mineral.

Calcium Fluoride

Oral fluids, such as saliva and the fluid phase of plaque (plaque fluid), are highly supersaturated with respect to CaF_2 after application of topical F agents, and hence this mineral has long been regarded as the primary source of bioavailable F in the oral environment [7–9]. CaF_2 dissolution and formation is governed by the same type of relationship noted for fluorapatite, with KSP_{CaF_2} $= 3 \times 10^{-10.4}$ [3] and $IAP_{CaF_2} = \{Ca^{2+}\} \times \{F^-\}^2$. Since resting oral fluids such as plaque fluid or saliva have a {Ca} of about 1 mmol/l [5, 10, 11], the critical value of free F that will induce a dissolution of pure CaF_2 can be calculated to be about 450 µmol/l [11]. This fluoride concentration is reached within about 10 min after a use of a conventional strength topical agent [12–14], and thus in theory CaF_2 deposits should rapidly dissolve in oral fluids within a short period after use of a topical F agent. Pure CaF_2 exposed to phosphate-containing solutions, such as saliva,

however has a very slow rate of dissolution due to adsorption of oral fluid phosphate onto the surface of the mineral [7, 15], probably in the form of fluorapatite [7]. These phosphate stabilized CaF_2 deposits lose PO_4^{-3} ions under low pH (cariogenic conditions) by protonation of these phosphate groups and thus dissolve rapidly [7, 9]. However CaF_2 formed in the presence of phosphate has considerably different properties from pure CaF_2, and some studies have suggested a moderately rapid rate of dissolution for the type of phosphate-containing CaF_2 deposits formed in the oral environment [8] while other studies have suggested that such deposits can be persistent [9, 16]. In this chapter, these phosphate-contaminated deposits are referred to, in accordance with other authors [7, 9], as 'calcium fluoride-like' (CaF_2-like).

Biologically/Bacterially Bound Fluoride

Bacterially Bound Fluoride
In an extensive study of bacterial Ca-F binding, Rose et al. [17] proposed a model (fig. 1) in which F reacts with intercellular or intracellular Ca 'bridges' to form calcium-fluoride (Ca-F) bonds at fixed anionic bacterial sites. An important feature of this model is that the application of F breaks bidentate Ca bonds, leading to more Ca and F binding (fig. 1A, B). These authors demonstrated that that bacterial Ca-F binding was quite unlike the binding of F by the mineral deposits described above: it is a *continuous* function of the $\{Ca^{2+}\}$ and $\{F^-\}$ which is also dependent on the number of binding sites on the bacteria (i.e. binding capacity) and the pH [17]. The pH dependence (fig. 1D) is a consequence of the competition of Ca^{2+} and H^+ for the same anionic sites on the bacterial surface.

Non-Bacterial Biological F Binding
Other 'biological' F binding sites exist in the oral environment besides bacteria – such as proteins,

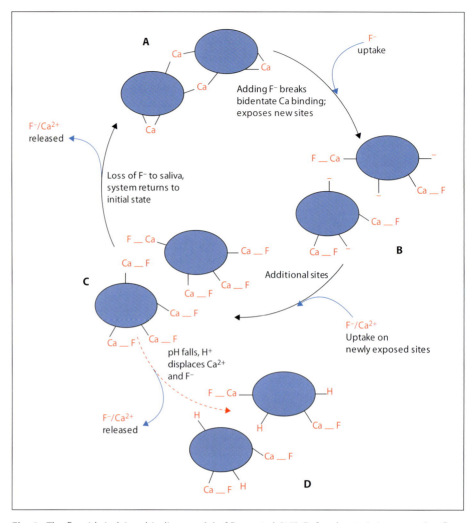

Fig. 1. The fluoride/calcium binding model of Rose et al. [17]. Before bacteria is exposed to fluoride (A), calcium (Ca) is intercellularly and intracellularly attached to bacteria in a bidentate fashion. Adding fluoride (F$^-$) breaks Ca bonds and exposes new binding sites (B), which leads to more F$^-$ and Ca^{2+} uptake (C). With time, F is lost to saliva and the bacteria returns to the initial state (A). However, when the bacteria is exposed to low pH, the increase in H$^+$ displaces Ca^{2+} and F$^-$ (D).

mucosal tissue and (most importantly with regard to plaque) bacterial fragments. It appears however that these moieties have binding properties that are similar to those described for bacteria in that they all appear to involve calcium-to-fluoride binding [11]. In this article these deposits are referred to generically as biological/bacterial Ca-F.

Location of Oral Bioavailable Fluoride Reservoirs

Bioavailable Fluoride Reservoirs on or in the Teeth

Fluorapatite
Due to the relative insolubility, and thus resistance to the acidic attack characteristic of caries,

the formation of fluorapatite (FAp) in/on the tooth had been considered as the primary goal of F therapies for many years [18]. Although a significant amount of FAp is found on the tooth surface (especially in caries-prone areas), this insoluble mineral is a poor source of oral fluid F. Thus, FAp provides little protection to adjacent F-poor tooth minerals, i.e. during caries progression, demineralization simply bypasses the FAp F-rich minerals in the outer layers of the lesion body and dissolves F-poor minerals at the advancing front.

CaF$_2$-Like Deposits

Many studies have identified CaF$_2$-like deposits as the most important labile source of F on/in the tooth surface [7, 9, 15, 16]. CaF$_2$-like deposits can be formed by the reaction of tooth-bound Ca with the applied F. However, at 'resting' pH, the low solubility of tooth mineral limits the rate of release of this ion. Thus, it appears doubtful that clinically significant amounts of this mineral could form on teeth [1, 6, 19] unless the pH of the topical agent is low (as in the case of APF [9, 20]), the application time is quite long (as with fluoride varnishes [19]), or a high concentration of the fluoride topical agents are employed [9, 16, 19]. Once formed, however, tooth deposits of this mineral have desirable properties as a F source, as evidenced by the cariostatic effect of infrequently applied APF rinses and varnishes [9]. The reasons for this beneficial effect include the location of these deposits at the site of the caries activity and their ability to release additional F during a cariogenic challenge [1, 7].

Quantification of Fluoride Reservoirs on or in the Teeth

Because CaF$_2$ is soluble in a basic low-phosphate solution, while FAp is nearly insoluble in such solutions, a sequential extraction of the tooth surface by base and then acid is often used to quantify CaF$_2$-like and FAp deposits on human teeth [16, 19]. For measuring small amounts of labile F, a constant composition procedure has been described, which not only quantifies these deposits, but also measures their rate of release into a 'saliva-like' solution of a chosen F concentration [21].

Bioavailable Fluoride Reservoirs in the Mucosa and Salivary Fluoride

Salivary Fluoride

During and shortly after administration of topical F agents, high levels of F are delivered to teeth and plaque via the saliva. However, salivary F concentrations rapidly fall below plaque fluid F levels [10, 12, 22], which tends to discount salivary F as a persistent source of plaque F. Salivary F, however, doubtlessly plays an important role in the remineralization of the plaque-free area of the teeth.

Location of Salivary F reservoirs

Zero et al. [23] found that edentulous subjects had higher salivary F levels than a dentate panel, suggesting that oral soft tissue is the major source of salivary F. Surprisingly, it has been shown that fluid recovered by scraping the mucosal surface after a F rinse is not only higher than saliva samples [24], but higher in F than plaque fluid samples recovered at the same time (unpublished data presented at the 1997 IADR meeting, abstract 174). Little is known of the nature of mucosal F reservoirs, other than they are easily depleted by water exposure, since post-water rinsing dramatically decreases post-topical application salivary F levels [25, 26]. However, in view of the fact that: (1) increases in salivary-free F appear to predict increases in plaque fluid F [5, 10, 12, 22, 27], and (2) salivary-free F and plaque fluid F appear to be correlated, at least in samples collected at times ≥ 2 h post application [12, 27], these deposits are likely to be similar in nature to those found in plaque.

Measurement of Whole and Free Salivary F

Although free saliva F, which is measured on centrifuged or filtered saliva, is more relevant to the cariostatic effect of F, salivary F is often reported on the whole saliva, which also includes F held in salivary particulates. Whole saliva F can be considerably larger than free saliva F, especially after use of Ca and phosphate-containing experimental agents [27].

Bioavailable Fluoride Reservoirs in Plaque and Plaque Fluid F

Plaque Fluid F

Changes in the concentration of ions in plaque fluid is the major factor governing the de/remineralization characteristic of caries [4, 5, 28, 29], and it is in this milieu that F exerts its major anti-caries effects [11, 29]. Thus 'bioavailable' plaque F reservoirs that substantively increase the plaque fluid F concentrations are of critical importance in the cariostatic effect of this ion.

Plaque F Reservoirs

Plaque fluid is greatly supersaturated with respect to FAp. Because FAp is sensitive to the high concentration of mineralization inhibitors found in the oral environment [30, 31], no significant stores of this mineral have been detected in plaque. More importantly, the insolubility of this mineral appears to negate its value as a plaque source of oral fluid F, and thus CaF_2-like deposits and biological/bacterial Ca-F appear to be the major plaque F reservoirs that increase plaque fluid F [7, 9, 17, 32, 33]. As described further below, recent studies [11] have suggested that unless additional Ca is supplied with conventional 'over the counter' topical agents, these reservoirs appear to be primarily in the form of biological/bacterial Ca-F deposits.

Relationship of Plaque and Plaque Fluid F

The relationship between plaque fluid and total plaque F in resting plaque is illustrated by the data in figure 2; samples were collected 1 h after a 228-µg/g F rinse given as NaMFP or NaF [22]. Here µg/g (ppm) refers to the mass fraction of F in the

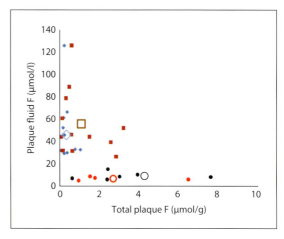

Fig. 2. Relationship between total plaque and plaque fluid F of samples recovered 1 h after a NaF (filled solid square), MFP (filled solid diamond) or 2 experimental rinses (filled circles) containing high levels of calcium and/or phosphate. The average values for these rinses are shown by the corresponding large open symbols. The individual MFP and NaF data points are unpublished data from reference 22. The red circle experimental rinse data is unpublished data from an abstract presented at the 2001 IADR meeting (abstract 1294) and refers to a controlled release rinse with a high Ca content (see text). The black circle data is unpublished data obtained by the author following the procedure of reference 22. This rinse contained high levels of both Ca and PO_4.

rinse. Although, there is no correlation between plaque fluid F and total plaque F, the average total plaque F after these 2 rinses reflects the average change in plaque fluid F. This is not always the case: the 2 experimental rinses shown in this figure deposited large amounts of plaque F, yet produced no increase in plaque fluid F. These rinses, which had high concentrations of Ca, and/or PO_4, appear to have formed primarily insoluble FAp.

Location of Plaque Fluoride Reservoirs, and the Effect of Water Rinsing on Plaque Fluoride

Studies performed on plaque samples formed in a 1-mm height ring and recovered after a topical F application have found a step gradient in total plaque F from the saliva to the enamel interface

[34]. In view of the tenuous relationship between total plaque and plaque fluid F noted above, the effect of this distribution on the subsequent distribution of plaque fluid F is unclear. However, given the shallow deposition of F found in these studies, the fact that a 30-second saliva-like wash administered after the F application induced no significant F loss suggests that the F deposits must not only form quickly, but release F slowly [34]. Noteworthy in this regard is a study that found only a small change in total plaque F in subjects who rinsed with water 1 h after use of a F rinse [10].

Release of Plaque F under Cariogenic Conditions
Studies on plaque fluid F, such as those described above, were performed on resting plaque, and thus are more relevant to the remineralization (rather than the demineralization) phase of the caries process. Unfortunately, studies of post-sucrose plaque can be complicated by several factors: (1) in high F plaque samples, F inhibits acid production [5, 35], raises the pH and reduces F release; (2) salivary clearance patterns maintain a high concentration of sucrose and F at the same sites, which increases F release from high F sites. As a result of these factors, studies in which plaque was recovered after a sucrose rinse often found no change or a decrease in plaque fluid F that was unrelated to the F content of the samples [5, 6, 28]. A better procedure for examining the F release from plaque F reservoirs at cariogenic pH appears to be the use of an in vitro acidification or titration [36].

Extraction Techniques for Examining the Properties of Plaque Fluoride Reservoirs
Water or buffer extraction has often been used to examine plaque reservoirs' ability to release F. Because of the difficulty in examining challenged plaque, such procedures are especially valuable when buffers are used whose pH is similar to cariogenic plaque [37]. Unfortunately, such methods are sensitive to the extraction conditions: given a large extraction volume and sufficient time even

the most insoluble F phases will completely dissolve (even at neutral pH). Such techniques are also sensitive to sample handling procedures, such as loss of fluid, which may induce conversion of these reservoirs to FAp.

Quantification of CaF$_2$-Like Deposits in Plaque
Recently, an extraction procedure was described that permits the quantification of CaF$_2$-like deposits in plaque recovered shortly after use of a F topical agent [11]: one of a pair of matched homogenized aliquots is repeatedly extracted with a very low phosphate-containing solution having the same $\{Ca^{2+}\}$, $\{F^-\}$ and $\{H^+\}$ as the plaque fluid recovered from these same samples. Since, shortly after a F rinse, plaque fluid IAP_{CaF_2} ($\{Ca^{2+}\} \times \{F^-\}^2$) is well below KSP_{CaF_2}, this extraction dissolves all the CaF$_2$-like deposits. However, plaque reservoirs that are in insoluble (FAp), or in equilibrium with, the 'plaque fluid-like' solution (biological/bacterial Ca-F) would not be extracted by this procedure. Hence, by comparing the total F content of this aliquot with the total F content of the unextracted aliquot, the amount of CaF$_2$-like deposits can be determined.

No CaF$_2$-Like Deposits Found in Plaque after Use of Over-the-Counter Strength Topical Agents
When the CaF$_2$ extraction procedure described above was applied to plaque samples recovered 30 min after a 228-µg/g F (NaF) rinse, the F content of the 2 aliquots was nearly identical [11]. This concentration of F delivered as a rinse appears to release more F to saliva and plaque than over-the-counter dentifrices [13]. Furthermore, the dentifrice ingredient sodium lauryl sulfate greatly reduces the formation of CaF$_2$-like deposits [7, 38]. Thus, these results suggest that biological/bacterial bound Ca-F, rather than CaF$_2$ or 'CaF$_2$-like' deposits, is the major reservoir of plaque F that releases this ion to plaque fluid in the case of over-the-counter F dentifrices and rinses. However, as described below, CaF$_2$-like deposits can be formed if additional Ca is supplied before a topical F agent.

Increasing the Deposition of Fluoride into Oral Reservoirs

Ca-F binding appears to play a central role in the formation of all the bioavailable oral F reservoirs; in fact nearly every study examining the relationship of Ca and F in plaque has found a moderate to strong correlation [33, 37–39]. Unfortunately, salivary and plaque fluid free-Ca is typically 2–10% of the amount of F supplied by topical agents. Thus, the amount of oral Ca-F that can be formed after use of a topical F agent is limited not only by the concentration of applied F, but also by the rate at which additional Ca can be scavenged from Ca reservoirs in enamel, plaque or saliva during the short period of F application [11, 37]. The small amounts of oral Ca-F reservoirs produced by this Ca scavenging explain why conventional topical agents induce only a transient increase in plaque fluid and salivary F before these reservoirs are exhausted [12, 13, 22, 33]. There are a number of procedures for increasing the formation of these Ca-F oral reservoirs, and consequently increasing the persistence of high levels of oral fluid F. Because such procedures may increase the cariostatic effect of a given F dose, they offer the possibility of decreasing the F content of topical agents without compromising the clinical effect.

Amorphous Calcium Phosphate Agents

Amorphous calcium phosphate products containing F appear to increase plaque F. A 450-μg/g F mouth rinse with casein phosphopeptide stabilized amorphous calcium phosphate was found to double the F content of plaque over a similar rinse without it [40]. However, it is difficult to separate the cariostatic effect of the F reservoirs produced by such products from the effects of the enhanced levels of Ca and phosphate. Furthermore, the inactivation of some of the applied F by formation of insoluble FAp is a concern with such products.

Two-Component Controlled Release Agents

Chow and colleagues [10, 12, 27, 36] have described a two-part 'controlled release' rinse in which part A contained Na_2SiF_6, while part B contained $CaCl_2$ and sodium acetate. This rinse initially also contained a low concentration of PO_4^{-3}; however, this component was eliminated in later studies. These rinses were called 'controlled release' (CR) agents because F was slowly released by the hydrolysis of the SiF_6^{2-} ion after parts A and B were mixed. The slow release of F in the presence of Ca could have two effects: (1) it permits an in-depth penetration by the SiF_6^{2-} and Ca ions into oral tissue and plaque before F release by hydrolysis of SiF_6^{2-} and subsequent formation of CaF_2 or biological/bacterial Ca-F deposits, or (2) alternatively, slowly growing nano-size crystals of CaF_2 could penetrate the above substrates before aggregation and growth.

Plaque and Salivary F after Use of the CR Rinse

Several studies have examined plaque F reservoirs 60 min after administration of a 228-μg/g F CR rinse. Compared to a NaF rinse with a similar F content, the CR rinse increased total plaque F by 4× [10], the amount of water extractable F by 11× [10], and low pH releasable F (average pH 5.2) by 9× [36]. Most importantly, the F reservoirs produced by the CR rinse appear to be bioavailable since, compared to the NaF rinse, they induce approximately a 2× increase in both plaque fluid and centrifuged saliva F [10, 12, 36]. Finally, the overnight F data of figure 3 show the persistence of these increases [27]. As noted above, the whole salivary F greatly exceeded the free (centrifuged) salivary F (fig. 3), especially in the Ca-containing CR rinses.

Enhanced Remineralization from CR Rinses

In lieu of a clinical trial, the best predictor of the cariostatic effectiveness of a topical F agent is a well-designed 'in situ de/remineralization' test protocol in which adequate positive and negative controls, reflecting the range of responses of known agents, is included [41]. The 228-μg/g F

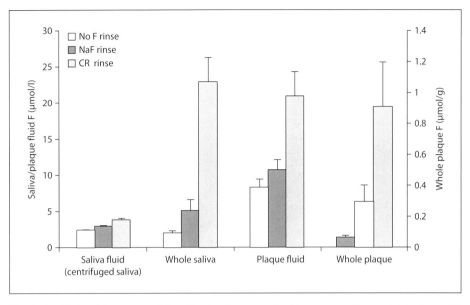

Fig. 3. Fluid fluoride and total fluoride in plaque and saliva measured overnight after a CR (controlled release, see text) or NaF rinse, both with 228 µg/g fluoride [27]. The error bars refer to standard error (n = 13). 'Whole' refers to the total amount of fluoride obtained by strong acid extraction of the sample, while 'fluid' refers to the fluoride in the supernatant of centrifuged samples. In all types of samples, CR rinse is significantly greater than the NaF rinse and no F rinse samples (p < 0.05).

CR rinse described above, when tested in such a procedure, was found to produce a remineralization effect that was not statistically different from a NaF rinse with 4× the F content [42]. Such results not only indicate the potential clinical effectiveness of the CR rinse, but also demonstrates clearly that the cariostatic effectiveness of any F product depends not just on the amount of applied F, but on the ability of the treatment to form bioavailable F deposits that substantively increase oral fluid F.

Problems with CR Agents
There are however problems with producing a commercially viable CR rinse or dentifrice: (1) some common dentifrice/rinse ingredients can reduce the deposition of F from CR agents; (2) although Na_2SiF_6 is approved for water fluoridation by the American Food and Drug Administration, it is not approved for use in topical agents. Chow

and colleagues however have described other NaF-based two-part CR systems that avoid some of these problems by using inhibitors (US patent No. 5891448) or Ca chelating agents (US patent No. 5476647) to control the rate of reaction of Ca with F.

Calcium Pre-Rinse Systems

A pre-application of a concentrated Ca agent shortly before a topical F agent is another procedure to ameliorate the restriction placed on the formation of oral Ca-F reservoirs by low oral fluid Ca.

Plaque and Salivary F after Use of a Calcium Pre-Rinse/NaF Rinse
Compared to the NaF rinse alone, Vogel et al. [37] found that total plaque F was elevated 12×, when

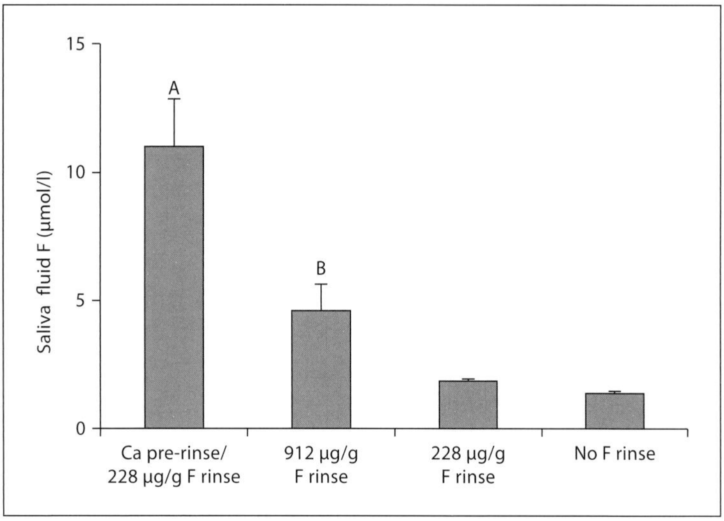

Fig. 4. Concentration of fluoride in centrifuged saliva samples obtained overnight after use of a 150-mmo/l calcium lactate pre-rinse immediately followed by a 228 µg/g F rinse; 228 µg/g F or 912 µg/g F (with no pre-rinse); or a distilled water rinse (no F rinse) [44]. All fluoride rinses were sodium fluoride. The error bars refer to the standard error (n = 12). Statistical differences are indicated by the letters.

a 150-mmol/l Ca pre rinse was used immediately before a 228-µg/g F (NaF) rinse. More importantly, the plaque fluid or centrifuged salivary F following the pre-rinse was about 5× greater than in the absence of the pre-rinse [37, 43]. These increases are persistent: in centrifuged saliva samples collected overnight after the above Ca pre-rinse/NaF rinse, F was 5.5× higher than after a NaF rinse alone, and 2.5× higher than after a rinse with 4 times more NaF (fig. 4) [44]. Given the relationship of salivary F and plaque fluid F noted above, it is noteworthy that the overnight increases in salivary F with the use of the Ca pre-rinse were much higher than found for the CR rinses (fig. 3), for which the more relevant total plaque F and plaque fluid F data are available.

Calcium Pre-Rinse Required to Produce Plaque CaF$_2$-Like Deposits from a Fluoride Rinse
An examination of plaque recovered 1 h after the use a Ca pre-rinse/NaF rinse using a variation of the CaF$_2$ extraction technique described above (unpublished data presented at the 2010 ORCA congress, abstract 75), found that, unlike the results obtained with the NaF rise alone, about one third of the deposited plaque F appeared to be in the form of CaF$_2$-like deposits. Given that the IAP$_{CaF_2}$ of saliva and plaque fluid immediately after use of topical F agents exceeds KSp$_{CaF_2}$ (supersaturation), the finding that a Ca pre-rinse is required to form CaF$_2$-like deposits may seem surprising. However, the IAP$_{CaF_2}$ ({Ca^{2+}} × {F$^-$}^{2}) required for the nucleation of this mineral is many orders of magnitude higher than KSP$_{CaF_2}$ [45], especially in a milieu such as plaque fluid that is rich in mineralization inhibitors, such as phosphate, which are known to reduce the rate of CaF$_2$ precipitation [46]. Furthermore, in the absence of the Ca pre-rinse, the low free Ca in plaque fluid or saliva implies that the formation of significant amounts of plaque CaF$_2$-like precipitates would require scavenging of additional Ca from the plaque (the small amount of CaF$_2$ formed in saliva would primarily migrate to the mucosa rather than into the plaque). Perhaps most importantly, this scavenging of plaque Ca by the applied F must compete with the formation of bacterial/biological Ca-F which occurs without such Ca extraction [17]. Hence it is unlikely, in the absence of the additional Ca supplied by a pre-rinse, that a

conventional topical F agent can form significant amounts of CaF_2-like deposits before {F} falls below the level required for deposition of this mineral. It is unclear however, if the large persistent increase in oral fluid F observed after use of the Ca pre-rinse/F rinse is due specifically to an increased formation of CaF_2-like deposits or to an increase in the total amount of F in oral reservoirs. The use of this extraction technique with samples recovered overnight should be able to address this question.

Problems with the Use of a Ca Pre-Application

There are a number of potential problems with using a Ca pre-application in conjunction with a F topical agent to increase oral F reservoirs: (1) a Ca-binding detergent found in many commercial F dentifrices, sodium lauryl sulfate, appears to reduce the ability of the Ca pre-application to increase salivary and plaque F [38, 43]; (2) patient compliance with a two-part regimen is problematic; however, it has been shown that Ca delivered as a dentifrice before a short F rinse (10 s) also produced large increases in salivary F [43]. Because such procedures are similar to current oral hygiene practices (substituting the F rinse for the usual post-dentifrice water rinse), they can potentially achieve a high level of compliance.

Conclusion

Current models for increasing the anti-caries effects of fluoride agents emphasize the importance of maintaining a cariostatic concentration of F in oral fluids. This has focused attention on the 'bioavailable' F reservoirs, dental plaque and the oral mucosa, which persistently increase the concentration of this ion in the fluid phases that are associated with the caries process, i.e. saliva and especially plaque fluid. Bioavailable F in these reservoirs appears to be of two types: 'CaF_2-like' reservoirs and fluoride bound to bacterial/biological substrates by Ca. Because fluoride binding in both these moieties appears to be mediated by Ca, their formation is limited by the low concentration of this ion in oral fluids. Several agents or procedures have been described that, by supplying additional Ca, greatly increase the deposition of Ca-F oral reservoirs and consequently increase the persistence of high levels of oral fluid fluoride. It is unclear, however, if such changes are related to the increases in the total amount of F in these reservoirs, or changes in the fraction of these deposits in the form of CaF_2-like or bacterial/biological reservoirs. New techniques have been developed for identifying these deposits which should prove useful in engineering topical agents that maximize the formation of oral fluoride reservoirs with optimum fluoride release characteristics. Such agents, by increasing the cariostatic effect of a given fluoride dose, offer the possibility of substantially decreasing the fluoride content of over-the-counter topical agents without compromising the clinical effect. Conversely, it appears possible to formulate conventional strength rinses and dentifrices that produce a considerably higher and more sustained F release into oral fluids than high-strength prescription products. Several studies have concluded that such agents do confer significant additional cariostatic effects [47–49].

References

1 ten Cate JM: Review on fluoride, with special emphasis on calcium fluoride mechanisms in caries prevention. Eur J Oral Sci 1997;105:461–465.

2 Featherstone JD: The science and practice of caries prevention. J Am Dent Assoc 2000;131:887–899.

3 McCann HG: The solubility of fluorapatite and its relationship to that of calcium fluoride. Arch Oral Biol 1968;13: 987–1001.

4   Vogel GL, Carey CM, Chow LC, Tatevossian A: Micro-analysis of plaque fluid from single-site fasted plaque. J Dent Res 1990;69:1316–1323.

5   Vogel GL, Zhang Z, Chow LC, Schumacher GE: Changes in lactate and other ions in plaque and saliva after a fluoride rinse and subsequent sucrose administration. Caries Res 2002;36:44–52.

6   Pearce ElF, Margolis H, Kent RL Jr: Effect of in situ plaque mineral supplementation on the state of saturation of plaque fluid during sugar induced acidogenesis. Eur J Oral Sci 1999;107:251–259.

7   Rølla G, Saxegaard E: Critical evaluation of the composition and use of topical fluorides with emphasis on the role of calcium fluoride in caries inhibition. J Dent Res 1990;69(Spec No):780–785.

8   Larsen MJ, Ravnholt G: Dissolution of various calcium fluoride preparations in inorganic solutions and in stimulated human saliva. Caries Res 1994;28:447–454.

9   Øgaard B: CaF$_2$ formation: cariostatic properties and factors of enhancing the effect. Caries Res 2001;35(suppl 1):40–44.

10 Vogel GL, Zhang Z, Chow LC, Schumacher GE: Effect of a water rinse on 'labile' fluoride and other ions in plaque and saliva before and after conventional and experimental fluoride rinses. Caries Res 2001;35:116–124.

11 Vogel GL, Tenuta LM, Schumacher GE, Chow LC: No calcium-fluoride-like deposits detected in plaque shortly after a sodium fluoride mouthrinse. Caries Res 2010;44:108–115.

12 Vogel GL, Mao Y, Carey CM, Chow LC, Takagi S: In vivo fluoride concentrations measured for two hours after a NaF or a novel two-solution rinse. J Dent Res 1992;71:448–452.

13 Zero DT, Raubertas RF, Fu J, Pedersen AM, Hayes AL, Featherstone JDB: Fluoride concentrations in plaque, whole saliva, and ductal saliva after application of home-use topical fluorides: J Dent Res 1992;71:1768–1775.

14 Bruun C, Givskov H, Thylstrup A: Whole saliva fluoride after toothbrushing NaF and MFP dentifrices with different F concentrations: Caries Res 1984;18:282–288.

15 Lagerlöf F, Saxegaard E, Barkvoll P, Rølla G: Effects of inorganic orthophosphate and pyrophosphate on dissolution of calcium fluoride in water. J Dent Res 1988;67:447–449.

16 Øgaard B, Rølla G, Helgeland K: Uptake and retention of alkali-soluble and alkali-insoluble fluoride in sound enamel in vivo after mouthrinses with 0.05% or 0.2% NaF. Caries Res 1983;17:520–524.

17 Rose RK, Shellis RP, Lee AR: The role of cation bridging in microbial fluoride binding: Caries Res 1996;30:458–464.

18 Chow LC: Tooth-bound fluoride and dental caries. J Dent Res 1990;69:595–600.

19 Cruz R, Øgaard B, Rølla G: Uptake of KOH-soluble and KOH-insoluble fluoride in sound human enamel after topical application of a fluoride varnish (Duraphat) or a neutral 2% NaF solution in vitro. Scand J Dent Res 1992;100:154–158.

20 Tsuda H, Jongebloed WL, Stokroos I, Arends J: Combined Raman and SEM study on CaF$_2$ formed on/in enamel by APF treatments. Caries Res 1993;27:445–454.

21 Sieck B, Takagi S, Chow LC: Assessment of loosely-bound and firmly-bound fluoride uptake by tooth enamel from topically applied fluoride treatments. J Dent Res 1990;69:1261–1265.

22 Vogel GL, Mao Y, Chow LC, Proskin HM: Fluoride in plaque fluid, plaque, and saliva measured for two hours after a NaF or NaMFP rinse. Caries Res 2000;34:404–411.

23 Zero DT, Raubertas RF, Pedersen AM, Fu J, Hayes AL, Featherstone JDB: Studies of fluoride retention by oral soft tissues after the application of home-use topical fluorides. J Dent Res 1992;71:1546–1552.

24 Jacobson AP, Stephen KW, Strang R: Fluoride uptake and clearance from the buccal mucosa following mouthrinsing. Caries Res 1992;26:56–58.

25 Duckworth RM, Knoop DT, Stephen KW: Effect of mouthrinsing after toothbrushing with a fluoride dentifrice on human salivary fluoride levels. Caries Res 1991;25:287–291.

26 Sjögren K, Birkhed D: Effect of various post-brushing activities on salivary fluoride concentration after toothbrushing with a sodium fluoride dentifrice: Caries Res 1994;28:127–131.

27 Vogel GL, Mao Y, Carey CM, Chow LC: Increased overnight fluoride concentrations in saliva, plaque, and plaque fluid after a two-solution rinse. J Dent Res 1997;76:761–767.

28 Tanaka M, Margolis HC: Release of mineral ions in dental plaque following acid production: Arch Oral Biol 1999:44:253–258.

29 Margolis HC, Moreno EC: Physicochemical perspectives on the cariostatic mechanisms of systemic and topical fluorides. J Dent Res 1990;69(Spec No):606–613.

30 Hay DI, Moreno EC: Macromolecular inhibitors of calcium phosphate precipitation in human saliva: their roles in providing a protective environment for the teeth; in Kleinberg I, Ellison SA, Mandel ID (eds): Saliva and Dental Caries. Microbiology Abstracts 1979;(sup suppl):45–58.

31 Tomazic B, Tomson M, Nancollas GH: Growth of calcium phosphates on hydroxyapatite crystals: the effect of magnesium. Arch Oral Biol 1975;20:803–808.

32 Larsen MJ, Richards A: The influence of saliva on the formation of calcium fluoride-like material on human dental enamel. Caries Res 2001;35:57–60.

33 Whitford GM, Wasdin JL, Schafer TE, Adair SM: Plaque fluoride concentrations are dependent on plaque calcium concentrations: Caries Res 2002;36:256–265.

34 Watson PS, Pontefract HA, Devine DA, Shore RC, Nattress BR, Kirkham J, Robinson C: Penetration of fluoride into natural plaque biofilms. J Dent Res 2005;84:451–455.

35 Hamilton IR: Biochemical effects of fluoride on oral bacteria. J Dent Res 1990;69(Special No):660–667.

36 Vogel GL, Zhang Z, Chow LC, Schumacher GE, Banting D: Effect of in vitro acidification on plaque fluid composition with and without a NaF or a controlled release fluoride rinse. J Dent Res 2000;79:983–990.

37 Vogel GL, Schumacher GE, Chow LC, Takagi S, Carey CM: Ca pre-rinse greatly increases plaque and plaque fluid F. J Dent Res 2008;87:466–469.

38 Pessan JP, Sicca CM, de Souza TS, da Silva SMB, Whitford GM, Buzalaf MAR: Fluoride concentrations in dental plaque and saliva after the use of a fluoride dentifrice preceded by a calcium lactate rinse. Eur J Oral Sci 2006;114:489–493.

39 Whitford GM, Buzalaf MAR, Bijella MFB, Waller JL: Plaque fluoride concentrations in a community without water fluoridation: effects of calcium and use of a fluoride or placebo dentifrice. Caries Res 2005;39:100–107.

40 Cochrane NJ, Cai F, Huq NL, Burrow MF, Reynolds EC: New approaches to enhanced remineralization of tooth enamel. J Dent Res 2010;89:1187–1197.

41 Proskin HM: Statistical considerations related to the use of caries model systems for the determination of clinical effectiveness of therapeutic agents. Adv Dent Res 1995;9:270–279.

42 Chow LC, Takagi S, Carey CM, Sieck BS: Remineralization effect of a two-solution fluoride mouth rinse – an in situ study. J Dent Res 2000;79:991–995.

43 Vogel GL, Shim D, Schumacher GE, Carey CM, Chow LC, Takagi S: Salivary fluoride from fluoride dentifrices or rinses after use of a calcium pre-rinse or calcium dentifrice. Caries Res 2006;40: 449–454.

44 Vogel GL, Chow LC, Carey CM: Calcium pre-rinse greatly increases overnight salivary fluoride after a 228 ppm fluoride rinse. Caries Res 2008;42:401–404.

45 Larsen MJ, Jensen SJ: Experiments on the initiation of calcium fluoride formation with reference to the solubility of dental enamel and brushite. Arch Oral Biol 1994;39:23–27.

46 Christoffersen J, Christoffersen MR, Kibalczyc W, Perdok WG: Kinetics of dissolution and growth of calcium fluoride and effects of phosphate. Acta Odontol Scand 1988;46:325–336.

47 Bartizek RD, Gerlach RW, Faller RV, Jacobs SA, Bollmer BW, Biesbrock AR: Reduction in dental caries with four concentrations of sodium fluoride in a dentifrice: a meta-analysis evaluation. J Clin Dent 2001;12:57–62.

48 Stookey GK, Mau MS, Isaacs RL, Gonzalez-Gierbolini C, Bartizek RD, Biesbrock AR: The relative anticaries effectiveness of three fluoride-containing dentifrices in Puerto Rico. Caries Res 2004;38: 542–550.

49 Tavss EA, Mellberg JR, Joziak M, Gambogi RJ, Fisher SW: Relationship between dentifrice fluoride concentration and clinical caries reduction. Am J Dent 2003;16:369–674.

Gerald Lee Vogel
American Dental Association Foundation, Paffenbarger Research Center
100 Bureau Drive Stop 8546, National Institute of Standards and Technology
Gaithersburg MD 20899–8546 (USA)
Tel. +1 301 975 6821, E-Mail jvogel@nist.gov

Buzalaf MAR (ed): Fluoride and the Oral Environment.
Monogr Oral Sci. Basel, Karger, 2011, vol 22, pp 158–170

Fluoride in Dental Erosion

Ana Carolina Magalhães[a] · Annette Wiegand[b] · Daniela Rios[a] ·
Marília Afonso Rabelo Buzalaf[a] · Adrian Lussi[c]

[a]Bauru Dental School, University of São Paulo, Bauru, Brazil; [b]University of Zürich, Zürich, and [c]University of Bern, Bern, Switzerland

Abstract

Dental erosion develops through chronic exposure to extrinsic/intrinsic acids with a low pH. Enamel erosion is characterized by a centripetal dissolution leaving a small demineralized zone behind. In contrast, erosive demineralization in dentin is more complex as the acid-induced mineral dissolution leads to the exposure of collagenous organic matrix, which hampers ion diffusion and, thus, reduces further progression of the lesion. Topical fluoridation inducing the formation of a protective layer on dental hard tissue, which is composed of CaF_2 (in case of conventional fluorides like amine fluoride or sodium fluoride) or of metal-rich surface precipitates (in case of titanium tetrafluoride or tin-containing fluoride products), appears to be most effective on enamel. In dentin, the preventive effect of fluorides is highly dependent on the presence of the organic matrix. In situ studies have shown a higher protective potential of fluoride in enamel compared to dentin, probably as the organic matrix is affected by enzymatical and chemical degradation as well as by abrasive influences in the clinical situation. There is convincing evidence that fluoride, in general, can strengthen teeth against erosive acid damage, and high-concentration fluoride agents and/or frequent applications are considered potentially effective approaches in preventing dental erosion. The use of tin-containing fluoride products might provide the best approach for effective prevention of dental erosion. Further properly designed in situ or clinical studies are recommended in order to better understand the relative differences in performance of the various fluoride agents and formulations.

Copyright © 2011 S. Karger AG, Basel

Dental erosion is defined as substance loss by exogenous or endogenous acids without bacterial involvement. The most important sources are dietary acids [1] and those originated from the stomach, like gastric acid from regurgitation and reflux disorders [2].

In contrast to initial caries, enamel erosion is predominantly a surface phenomenon with a centripetal bulk substance loss combined with a small partly demineralized surface layer with decreased microhardness (fig. 1). In dentin, the erosive demineralization is mostly diffusion controlled, as the increasing exposure of organic matrix hampers ion diffusion, and thus reduces further progression of dentin erosion (fig. 2) [3, 4].

A.C.M. and A.W. contributed equally.

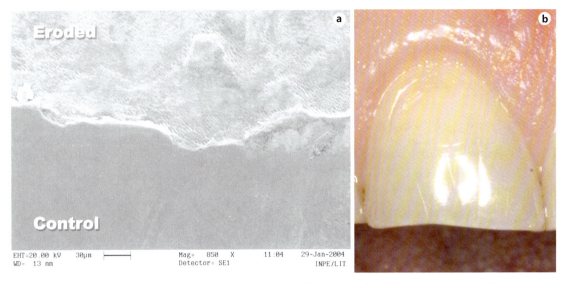

Fig. 1. Scanning electron microscopy (**a**) and clinical picture (**b**) of enamel erosion. Images are not from the same tooth.

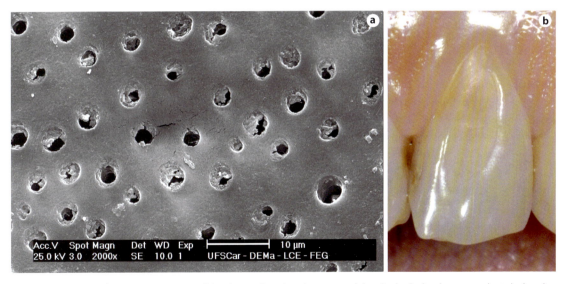

Fig. 2. a Scanning electron microscopy of dentin erosion showing opened dentinal tubules; however, the tubules also can be partially or totally closed in the clinical situation. Reprinted from Kato et al. [59], with permission. **b** Clinical picture of dentin erosion. Images are not from the same tooth.

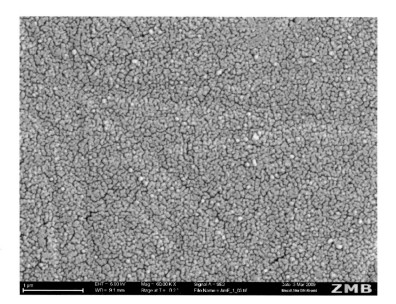

Fig. 3. Scanning electron microscopy of enamel treated with conventional fluoride (AmF, 0.5 M fluoride, pH 4.5, applied for 60 s).

There is evidence that the prevalence of erosion is steadily increasing [5]. Preventive strategies in the management of dental erosion consider dietary counseling, stimulation of salivary flow, modification of erosive beverages, adequate oral hygiene measures and fluoride treatment as the most relevant [6].

This chapter will give an overview of the current knowledge on the use of fluorides, including conventional and metal fluorides, for the prevention of erosive and combined erosive-abrasive dental loss. Due to the fact that the histology of enamel and dentin erosion is considerably different, this chapter will be divided into two parts: (1) fluorides and enamel erosion; (2) fluorides and dentin erosion.

Fluorides and Enamel Erosion

Extrinsic and/or intrinsic acids with low pH (pH 1.0–3.5) initially cause either the dissolution of the prism cores or interprismatic areas, showing a honeycomb structure in prismatic enamel. In aprismatic enamel, the demineralization is irregular, without a clear structural pattern. If the erosive challenge is ongoing, the dissolution process results in surface loss accompanied by a progressive softening of the surface. As the demineralized layer of eroded enamel is considerably small when compared to the enamel loss, fluoride application predominately aims to prevent erosive tissue loss rather than to remineralize softened enamel.

Conventional fluorides, whose beneficial effect against caries is well known [7], have been tested for prevention or control of dental erosion [8]. The potential of conventional fluorides, such as sodium fluoride (NaF) and amine fluoride (AmF), to prevent erosive demineralization is mainly related to the formation of a calcium fluoride (CaF_2) layer [9, 10] (fig. 3). This layer is assumed to behave as a physical barrier that hampers the contact of the acid with the underlying enamel or to act as a mineral reservoir which is attacked by the erosive challenge. Thereafter, released calcium and fluoride might increase the saturation level with respect

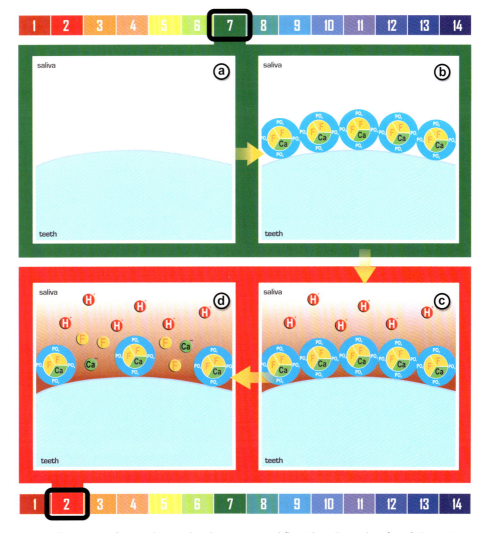

Fig. 4. Illustration of enamel treated with conventional fluoride. **a** Enamel surface. **b** Deposition of a CaF_2 layer. **c** CaF_2 layer acting as a physical barrier for the erosive challenge. **d** Progressive CaF_2 layer dissolution.

to dental hard tissue in the liquid adjacent to the surface, thus promoting remineralization (fig. 4, 5).

The formation of the CaF_2-like layer and its protective effect against demineralization is highly dependent on the pH, the concentration of fluoride and the frequency of application. The deposition of CaF_2 on the surface increases with increasing concentration and frequency of application and decreasing pH of the agent. Fluoride agents with a pH below 5 seem to induce a higher CaF_2 deposition on dental surface than neutral ones [9].

Ganss et al. [10] evaluated the retention of CaF_2 on human enamel under neutral and acidic conditions in vitro and in situ. Fluoride

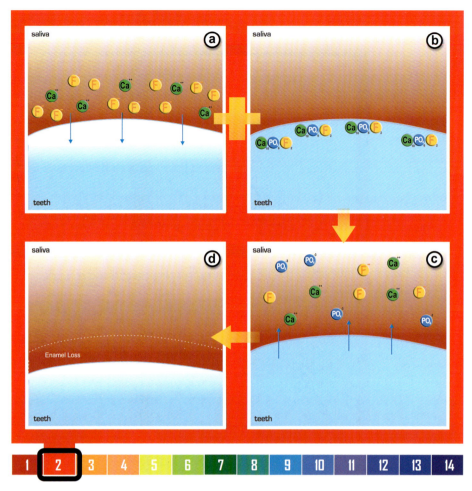

Fig. 5. Illustration of enamel treated with conventional fluoride. **a** CaF_2 layer final dissolution. **b** Simultaneous calcium and fluoride saturation provoking remineralization. **c** Subsequent erosive challenge. **d** Bulk substance loss combined with a small partly demineralized surface layer.

(10,000 ppm, AmF) was applied once for 5 min, and the enamel specimens were exposed to erosive demineralization (3 × 30 s/day, 4 days in vitro; 3 × 2 min/day, 7 days in situ) or neutral conditions (artificial saliva in vitro; human saliva in situ). It was shown that more CaF_2 was lost under erosive compared to neutral conditions in vitro, while the intraoral environment was considerably protective for CaF_2-like precipitates, especially on enamel.

Although toothbrushing might affect the progression of eroded dental hard tissues adversely by removing the softened layer of enamel [11, 12], it was shown that the use of fluoridated (NaF) toothpastes might diminish the abrasive effect to some extent [13–15]. However, as the overall protective effect of toothpastes with 1,100–5,000 µg/g fluoride is limited [14, 15], the use of highly concentrated fluoride varnishes (22,600 µg/g) was anticipated to be more effective due to their capacity

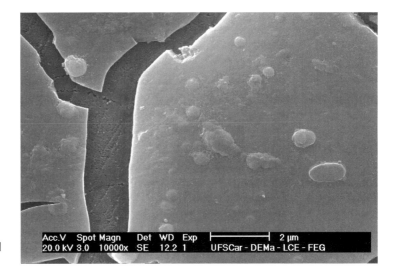

Fig. 6. Scanning electron microscopy of enamel treated with 4% titanium tetrafluoride varnish (6 h). Reprinted from Magalhães et al. [19] with permission.

Acc.V Spot Magn Det WD Exp | 2 µm
20.0 kV 3.0 10000x SE 12.2 1 | UFSCar - DEMa - LCE - FEG

to adhere to the tooth surface and create a CaF_2 reservoir [16, 17]. Indeed, the application of NaF varnish (22,600 µg/g) was effective in reducing enamel erosion for 30 min of acid exposure, but the protective effect declined thereafter [18, 19]. However, as placebo varnishes also showed some protection against enamel erosion and combined erosion/abrasion, it is believed that the protective effect of fluoride varnishes is mainly related to the mechanical rather than to the chemical protection [20, 21].

As the anti-erosive effect of conventional fluorides requires a very intensive fluoridation regime [22], recent studies have focused on fluoride compounds which might deliver a higher level of efficacy. In this context, compounds containing polyvalent metal ions such as stannous fluoride or titanium tetrafluoride were tested.

Several in vitro studies have shown an inhibitory effect of 0.4–10% TiF_4 solution on dental erosion [23–27], which is attributed not only to the effect of fluoride, but mainly to the action of titanium [23, 28]. Its protective effect is related to the formation of an acid-resistant surface coating, the increased fluoride uptake and

the titanium incorporation in the hydroxyapatite lattice. The glaze-like surface layer observed after the application of TiF_4 is assumed to be due to the formation of a new compound (hydrated hydrogen titanium phosphate) that might primarily act as a diffusion barrier [23, 29–32] (fig. 6, 7). The increased fluoride uptake found after application of TiF_4 can be explained by the ability of the polyvalent metal ion to form strong fluoride complexes firmly bound to the apatite crystals [30, 32].

Information regarding the efficacy of TiF_4 under clinical conditions is scarce and contradictory, as only two in situ studies showed 1.6% TiF_4 (0.5 M fluoride) to be as effective as SnF_2 or AmF in the prevention of erosion or combined erosion/abrasion [33, 34], while other did not show any protective effect of 4% TiF_4 [20, 21, 35]. The efficacy of TiF_4 is highly dependent on the pH of the agent, since it was shown that enamel erosion can be significantly reduced by TiF_4 (0.5 M fluoride) at native pH (pH 1.2) but not at a pH buffered to 3.5 [36]. One study indicated that TiF_4 applied in the form of a varnish might be of higher efficacy than as a solution [19]. However, it should be

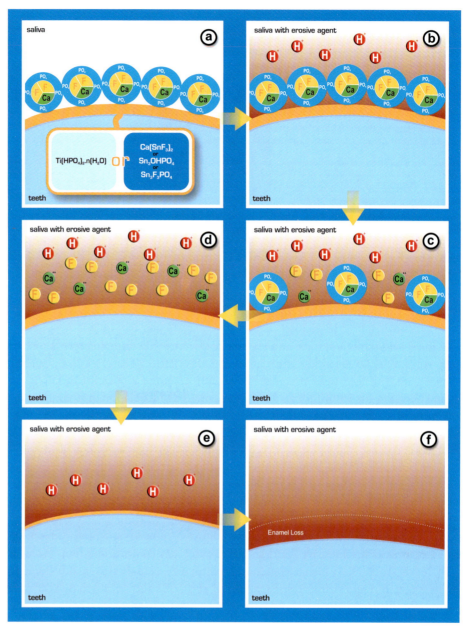

Fig. 7. Illustration of the formation of (CaF$_2$) layer and an acid-resistant surface coating composed of hydrated hydrogen titanium phosphate after the application of TiF$_4$, or composed of metal-rich precipitates [Ca(SnF$_3$)$_2$, SnOHPO$_4$, Sn$_3$F$_3$PO$_4$] after the application of tin-containing fluoride mouth rinses. **a** CaF$_2$ layer and the metal-rich precipitates (in orange). **b** Erosive challenge. **c** CaF$_2$ layer dissolution. **d** CaF$_2$ layer final dissolution and the preservation of the metal-rich precipitate. **e** Progressive erosive challenges. **f** Final dissolution of the metal-rich layer and consequent enamel loss.

Magalhães · Wiegand · Rios · Buzalaf · Lussi

Fig. 8. Scanning electron microscopy of enamel treated with SnF_2 solution (0.48 M, pH 2.7, 3 min) before erosion. Reprinted from Yu et al. [60] with permission.

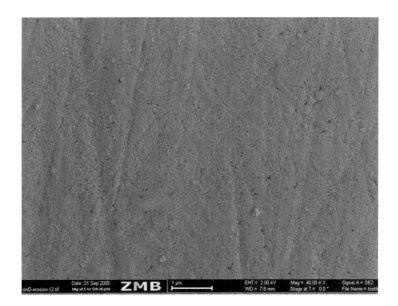

Fig. 9. Scanning electron microscopy of enamel treated with SnF_2 solution after erosion (6 × 1 min/ day, 5 days), showing no alteration. Reprinted from Yu et al. [60] with permission.

considered that the low pH of TiF_4 products does not allow self-application by the patient.

Tin-containing fluoride products have shown promising results in several studies [37–41]. The mode of action of tin-containing fluoride solutions is probably attributed to the formation of metal-rich surface precipitates [$Ca(SnF_3)_2$, $SnOHPO_4$, $Sn_3F_3PO_4$], which were shown to be of high acid resistance [42] (fig. 7–9). Further, tin may penetrate and become incorporated into the demineralized layer when high concentrated tin containing fluoride mouth rinses are used [38, 43].

Ganss et al. [44] evaluated the relevance of cations in different fluoride compounds for their effectiveness as anti-erosive agents and showed that $SnCl_2$ (800 ppm tin), NaF (250 ppm fluoride), AmF/SnF_2 (250 ppm fluoride/390 ppm tin) and SnF_2 (250 ppm fluoride/809 ppm tin) solutions could reduce enamel erosion. Treatment with solutions containing SnF_2 was most effective. The combination of $AmF/NaF/SnCl_2$ with high (2,800 ppm tin/1,500 ppm fluoride) and low (700 ppm Sn/1,500 ppm fluoride) tin concentrations reduced erosion by 90 and 70%, respectively [38, 39].

Some possible side effects of high-concentration tin-containing mouth rinses may include a dull feeling on the tooth surface, astringent sensation and tooth discoloration (1,900 ppm tin) [45]. Therefore, tin-containing solutions of lower concentration (800 ppm tin/ 500 ppm fluoride) were tested in vitro and in situ [46, 47]. Under severe erosive conditions, the $SnCl_2/NaF/AmF$ exhibited a high potential to reduce enamel erosion (67% reduction), and showed no adverse side effects [47]. Besides mouth rinses, tin-containing fluoride toothpastes were tested using in vitro protocols, and shown to perform significantly better under erosive challenges when compared with NaF- and MFP-containing toothpastes [41]. Further research should test specially formulated tin-containing fluoride products to minimize aesthetic negatives seen with high-concentration tin-containing products, which may provide a highly effective means to help prevent dental erosion using a consumer-friendly approach.

Fluorides and Dentin Erosion

The preventive effect of fluorides on dentin erosion is highly dependent on the presence of the organic matrix [48]. Initial studies showed that a very intensive fluoridation regimen combining toothpaste (0.15% fluoride, NaF), mouth rinse (0.025% fluoride, AmF/NaF) and gel (1.25% fluoride, AmF/NaF) was most effective in the prevention of dentin erosion [22, 49]. However, after enzymatic removal of the organic matrix, fluoride was ineffective [3, 50]. It was assumed that the demineralized organic dentin matrix has a buffering capacity sufficient to prevent further dentin demineralization, especially in the presence of high amounts of fluoride [3]. Moreover, the exposed organic matrix of etched dentin involves an increased surface area and increased diffusion pathways – enhancing the amount of structurally bound and KOH-soluble fluoride compared to sound dentin [51]. However, it remains unclear to what extent the organic material is retained under clinical conditions, when the collagen layer might be affected by enzymatical and chemical degradation [50, 52] as well as by abrasive influences. From the clinical appearance of dentin-erosive lesions, it seems likely that the collagenous layer is at least partly removed. This hypothesis might also explain why fluorides such as NaF were less effective in dentin than in enamel under in situ conditions [10, 22, 38] but not in laboratory experiments [27, 53].

The application of slightly acidic fluoride formulations such as NaF or AmF results in the formation of CaF_2 precipitates on both enamel and dentin (fig. 10), but the precipitates are less stable on dentin than on enamel under erosive conditions [10]. Although the preventive potential of NaF and AmF solution and dentifrice on dentin erosion and combined erosion/abrasion was shown in different in situ studies [22, 34, 54], information about the ideal fluoride concentration and frequency of application is scarce. Also, the resistance of dentinal CaF_2 precipitates against abrasion has not so far been assessed directly; only an in situ study indicated that the protective potential of AmF against erosion is not affected by additional brushing treatment [34].

Considering the severe and chronic acid exposure in patients suffering from dental erosion, the effect of CaF_2 precipitates is probably limited

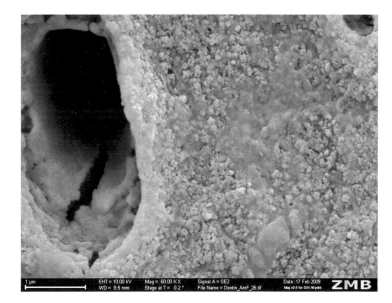

Fig. 10. Scanning electron microscopy of dentin treated with conventional fluoride (AmF, 0.5 M F, pH 4.5, applied for 60 s).

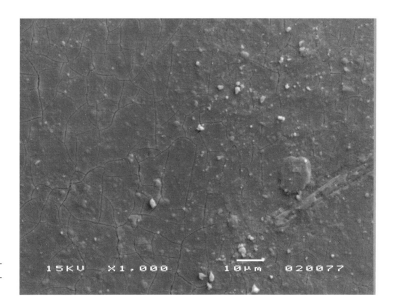

Fig. 11. Scanning electron microscopy of dentin treated with 4% titanium tetrafluoride varnish (6 h).

over time [10], and fluoride compounds with a distinct potential to resist an erosive challenge are required.

Titanium tetrafluoride was shown to induce some coating on dentin surfaces, which partly covered dentinal tubules [55] (fig. 11). However, its protective potential did not exceed the efficacy of NaF or AmF [27, 34, 56], and the low pH required for the efficacy of the agents has not so far allowed for a clinical application [57].

Tin-containing fluoride solutions have been demonstrated to exhibit promising anti-erosive effects not only on enamel but also on dentin [38, 44, 46]. The suggested mechanism of action is related to the incorporation of tin in mineralized dentin when the organic matrix is allowed to develop and to surface precipitation when the organic matrix is enzymatically removed [58]. In cases where the organic matrix is preserved, phosphorus, phosphorylated phosphoprotein or phosphophoryn might attract the tin ion, which is then retained in the organic matrix to some extent, but also accumulates in the underlying mineralized tissue. In cases where the organic matrix is removed, tin reacts with the mineral by forming different salts, e.g. $Sn(OH)_2$, $Sn_2(PO_4)OH$, $Ca(SnF_3)$, $Sn_3F_3PO_4$, $Sn_2(OH)PO_4$, $Sn_3F_3PO_4$ or $SnHPO_4$ [58]. Recent in situ studies demonstrated that mouth rinses containing $AmF/NaF/SnCl_2$ (500 ppm F, 800 ppm Sn) reduced dentin erosion by 50% and were significantly more effective than an NaF-containing mouth rinse (500 ppm F) [38, 47].

Comparing the protective effect of different fluoride compounds on dentin erosion, Ganss et al. [48] showed that solutions containing AmF and/or SnF_2 performed only slightly better than solutions containing NaF and/or AmF in the presence of the organic matrix. However, continuous removal of the organic matrix influenced the efficacy of the fluoride compounds distinctly and demonstrated a significantly better preventive effect of the SnF_2- and AmF/SnF_2-containing solutions compared to all other solutions.

Conclusion

Conventional fluorides with a known anti-cariogenic potential offer some, but limited, protection against erosion as the CaF_2 precipitates formed on the surface are readily soluble in acids. Metal-containing fluoride compounds showed promising results in prevention of erosion, but might involve some adverse side effects due to the very low pH (in case of titanium tetrafluoride) and the potential to cause slight discoloration, a dull feeling on the tooth surface and an astringent sensation (in case of highly concentrated tin-containing fluoride solutions).

There is convincing evidence that fluoride, in general, can strengthen enamel against erosive acid damage; high-concentration fluoride agents and/or frequent applications are considered potentially effective approaches to prevent dental erosion. However, fluorides might be more effective in enamel than in dentin, as the organic matrix influencing the efficacy of fluorides might to some extent be affected by enzymatical and chemical degradation as well as by mechanical abrasion. The use of tin-containing fluoride products might provide the best approach for effective prevention of dental erosion.

References

1 Lussi A, Jaeggi T, Zero D: The role of diet in the aetiology of dental erosion. Caries Res 2004;38(suppl 1):34–44.

2 Bartlett D: Intrinsic causes of erosion. Monogr Oral Sci 2006;20:119–139.

3 Ganss C, Klimek J, Starck C: Quantitative analysis of the impact of the organic matrix on the fluoride effect on erosion progression in human dentine using longitudinal microradiography. Arch Oral Biol 2004;49:931–935.

4 Hara AT, Ando M, Cury JA, Serra MC, Gonzalez-Cabezas C, Zero DT: Influence of the organic matrix on root dentine erosion by citric acid. Caries Res 2005; 39:134–138.

5 Lussi A: Erosive tooth wear – a multifactorial condition of growing concern and increasing knowledge. Monogr Oral Sci 2006;20:1–8.

6 Magalhães AC, Wiegand A, Rios D, Honorio HM, Buzalaf MA: Insights into preventive measures for dental erosion. J Appl Oral Sci 2009;17:75–86.

7 ten Cate JM: Review on fluoride, with special emphasis on calcium fluoride mechanisms in caries prevention. Eur J Oral Sci 1997;105:461–465.

8 Wiegand A, Attin T: Influence of fluoride on the prevention of erosive lesions – a review. Oral Health Prev Dent 2003; 1:245–253.

9 Saxegaard E, Rolla G: Fluoride acquisition on and in human enamel during topical application in vitro. Scand J Dent Res 1988;96:523–535.

10 Ganss C, Schlueter N, Klimek J: Retention of KOH-soluble fluoride on enamel and dentine under erosive conditions – a comparison of in vitro and in situ results. Arch Oral Biol 2007;52:9–14.

11 Attin T, Knofel S, Buchalla W, Tutuncu R: In situ evaluation of different remineralization periods to decrease brushing abrasion of demineralized enamel. Caries Res 2001;35:216–222.

12 Rios D, Honorio HM, Magalhães AC, Delbem AC, Machado MA, Silva SM, Buzalaf MA: Effect of salivary stimulation on erosion of human and bovine enamel subjected or not to subsequent abrasion: an in situ/ex vivo study. Caries Res 2006;40:218–223.

13 Bartlett DW, Smith BG, Wilson RF: Comparison of the effect of fluoride and non-fluoride toothpaste on tooth wear in vitro and the influence of enamel fluoride concentration and hardness of enamel. Br Dent J 1994;176:346–348.

14 Magalhães AC, Rios D, Delbem AC, Buzalaf MA, Machado MA: Influence of fluoride dentifrice on brushing abrasion of eroded human enamel: an in situ/ex vivo study. Caries Res 2007;41:77–79.

15 Rios D, Magalhães AC, Polo RO, Wiegand A, Attin T, Buzalaf MA: The efficacy of a highly concentrated fluoride dentifrice on bovine enamel subjected to erosion and abrasion. J Am Dent Assoc 2008;139:1652–1656.

16 Sorvari R, Meurman JH, Alakuijala P, Frank RM: Effect of fluoride varnish and solution on enamel erosion in vitro. Caries Res 1994;28:227–232.

17 Vieira A, Jager DH, Ruben JL, Huysmans MC: Inhibition of erosive wear by fluoride varnish. Caries Res 2007;41:61–67.

18 Magalhães AC, Stancari FH, Rios D, Buzalaf MA: Effect of an experimental 4% titanium tetrafluoride varnish on dental erosion by a soft drink. J Dent 2007;35:858–861.

19 Magalhães AC, Kato MT, Rios D, Wiegand A, Attin T, Buzalaf MA: The effect of an experimental 4% Tif4 varnish compared to NaF varnishes and 4% TiF4 solution on dental erosion in vitro. Caries Res 2008;42:269–274.

20 Vieira A, Ruben JL, Huysmans MC: Effect of titanium tetrafluoride, amine fluoride and fluoride varnish on enamel erosion in vitro. Caries Res 2005;39:371–379.

21 Vieira A, Lugtenborg M, Ruben JL, Huysmans MC: Brushing abrasion of eroded bovine enamel pretreated with topical fluorides. Caries Res 2006;40:224–230.

22 Ganss C, Klimek J, Brune V, Schurmann A: Effects of two fluoridation measures on erosion progression in human enamel and dentine in situ. Caries Res 2004;38:561–566.

23 Buyukyilmaz T, Ogaard B, Rolla G: The resistance of titanium tetrafluoride-treated human enamel to strong hydrochloric acid. Eur J Oral Sci 1997;105:473–477.

24 van Rijkom H, Ruben J, Vieira A, Huysmans MC, Truin GJ, Mulder J: Erosion-inhibiting effect of sodium fluoride and titanium tetrafluoride treatment in vitro. Eur J Oral Sci 2003;111:253–257.

25 Hove L, Holme B, Ogaard B, Willumsen T, Tveit AB: The protective effect of TiF₄, SnF₂ and NaF on erosion of enamel by hydrochloric acid in vitro measured by white light interferometry. Caries Res 2006;40:440–443.

26 Hove LH, Young A, Tveit AB: An in vitro study on the effect of TiF(4) treatment against erosion by hydrochloric acid on pellicle-covered enamel. Caries Res 2007;41:80–84.

27 Schlueter N, Ganss C, Mueller U, Klimek J: Effect of titanium tetrafluoride and sodium fluoride on erosion progression in enamel and dentine in vitro. Caries Res 2007;41:141–145.

28 Tezel H, Ergucu Z, Onal B: Effects of topical fluoride agents on artificial enamel lesion formation in vitro. Quintessence Int 2002;33:347–352.

29 Mundorff SA, Little MF, Bibby BG: Enamel dissolution. II. Action of titanium tetrafluoride. J Dent Res 1972;51:1567–1571.

30 Wei SH, Soboroff DM, Wefel JS: Effects of titanium tetrafluoride on human enamel. J Dent Res 1976;55:426–431.

31 Ribeiro CC, Gibson I, Barbosa MA: The uptake of titanium ions by hydroxyapatite particles-structural changes and possible mechanisms. Biomaterials 2006;27:1749–1761.

32 Wiegand A, Magalhães AC, Attin T: Is titanium tetrafluoride (TiF4) effective to prevent carious and erosive lesions? A review of the literature. Oral Health Prev Dent 2010;8:159–164.

33 Hove LH, Holme B, Young A, Tveit AB: The protective effect of TiF₄, SnF₂ and NaF against erosion-like lesions in situ. Caries Res 2008;42:68–72.

34 Wiegand A, Hiestand B, Sener B, Magalhães AC, Roos M, Attin T: Effect of TiF(4), ZrF(4), HfF(4) and AmF on erosion and erosion/abrasion of enamel and dentin in situ. Arch Oral Biol 2010;55:223–228.

35 Magalhães AC, Rios D, Honorio HM, Jorge AM Jr, Delbem AC, Buzalaf MA: Effect of 4% titanium tetrafluoride solution on dental erosion by a soft drink: an in situ/ex vivo study. Arch Oral Biol 2008;53:399–404.

36 Wiegand A, Waldheim E, Sener B, Magalhães AC, Attin T: Comparison of the effects of TiF₄ and NaF solutions at pH 1.2 and 3.5 on enamel erosion in vitro. Caries Res 2009;43:269–277.

37 Schlueter N, Klimek J, Ganss C: Effect of stannous and fluoride concentration in a mouth rinse on erosive tissue loss in enamel in vitro. Arch Oral Biol 2009;54:432–436.

38 Schlueter N, Klimek J, Ganss C: Efficacy of an experimental tin-F-containing solution in erosive tissue loss in enamel and dentine in situ. Caries Res 2009;43:415–421.

39 Schlueter N, Klimek J, Ganss C: In vitro efficacy of experimental tin- and fluoride-containing mouth rinses as anti-erosive agents in enamel. J Dent 2009;37:944–948.

40 Hooper SM, Newcombe RG, Faller R, Eversole S, Addy M, West NX: The protective effects of toothpaste against erosion by orange juice: studies in situ and in vitro. J Dent 2007;35:476–481.

41 Faller RV, Eversole SL, Tzeghai GE: Enamel Protection: A Comparison of Marketed Dentifrice Performance Against Dental Erosion. Am J Dent 2011, in press.

42 Babcock FD, King JC, Jordan TH: The reaction of stannous fluoride and hydroxyapatite. J Dent Res 1978;57:933–938.

43 Schlueter N, Hardt M, Lussi A, Engelmann F, Klimek J, Ganss C: Tin-containing fluoride solutions as anti-erosive agents in enamel: an in vitro tin-uptake, tissue-loss, and scanning electron micrograph study. Eur J Oral Sci 2009;117:427–434.

44 Ganss C, Schlueter N, Hardt M, Schattenberg P, Klimek J: Effect of fluoride compounds on enamel erosion in vitro: a comparison of amine, sodium and stannous fluoride. Caries Res 2008;42: 2–7.

45 Schlueter N, Klimek J, Ganss C: Efficacy of tin-containing solutions on erosive mineral loss in enamel and dentine in situ. Clin Oral Investig 2010, E-pub ahead of print.

46 Schlueter N, Neutard L, von Hinckeldey J, Klimek J, Ganss C: Tin and fluoride as anti-erosive agents in enamel and dentine in vitro. Acta Odontol Scand 2010, E-pub ahead of print.

47 Ganss C, Neutard L, von Hinckeldey J, Klimek J, Schlueter N: Efficacy of a tin/fluoride rinse: a randomized in situ trial on erosion. J Dent Res 2010;89: 1214–1218.

48 Ganss C, Lussi A, Sommer N, Klimek J, Schlueter N: Efficacy of fluoride compounds and stannous chloride as erosion inhibitors in dentine. Caries Res 2010; 44:248–252.

49 Ganss C, Klimek J, Schaffer U, Spall T: Effectiveness of two fluoridation measures on erosion progression in human enamel and dentine in vitro. Caries Res 2001;35:325–330.

50 Schlueter N, Ganss C, Hardt M, Schegietz D, Klimek J: Effect of pepsin on erosive tissue loss and the efficacy of fluoridation measures in dentine in vitro. Acta Odontol Scand 2007;65:298–305.

51 Buchalla W, Lennon AM, Becker K, Lucke T, Attin T: Smear layer and surface state affect dentin fluoride uptake. Arch Oral Biol 2007;52:932–937.

52 Schlueter N, Hardt M, Klimek J, Ganss C: Influence of the digestive enzymes trypsin and pepsin in vitro on the progression of erosion in dentine. Arch Oral Biol 2010;55:294–299.

53 Wiegand A, Meier W, Sutter E, Magalhães AC, Becker K, Roos M, Attin T: Protective effect of different tetrafluorides on erosion of pellicle-free and pellicle-covered enamel and dentine. Caries Res 2008;42:247–254.

54 Magalhães AC, Rios D, Moino AL, Wiegand A, Attin T, Buzalaf MA: Effect of different concentrations of fluoride in dentifrices on dentin erosion subjected or not to abrasion in situ/ex vivo. Caries Res 2008;42:112–116.

55 Wiegand A, Magalhães AC, Navarro RS, Schmidlin PR, Rios D, Buzalaf MA, Attin T: Effect of titanium tetrafluoride and amine fluoride treatment combined with carbon dioxide laser irradiation on enamel and dentin erosion. Photomed Laser Surg 2010;28:219–226.

56 Magalhães AC, Levy FM, Rios D, Buzalaf MA: Effect of a single application of TiF(4) and NaF varnishes and solutions on dentin erosion in vitro. J Dent 2010; 38:153–157.

57 Wiegand A, Magalhães AC, Sener B, Waldheim E, Attin T: TiF(4) and NaF at pH 1.2 but not at pH 3.5 are able to reduce dentin erosion. Arch Oral Biol 2009;54:790–795.

58 Ganss C, Hardt M, Lussi A, Cocks AK, Klimek J, Schlueter N: Mechanism of action of tin-containing fluoride solutions as anti-erosive agents in dentine – an in vitro tin-uptake, tissue loss, and scanning electron microscopy study. Eur J Oral Sci 2010;118:376–384.

59 Kato MT, Leite AL, Hannas AR, Oliveira RC, Pereira JC, Tjäderhane L, Buzalaf MA: Effect of iron on matrix metalloproteinase inhibition and on the prevention of dentine erosion. Caries Res 2010;44:309–316.

60 Yu H, Attin T, Wiegand A, Buchalla W: Effects of various fluoride solutions on enamel erosion in vitro. Caries Res 2010;44:390-401.

Prof. Ana Carolina Magalhães
Bauru Dental School, University of São Paulo
Al. Dr. Octávio Pinheiro Brisolla 9–75
Bauru-SP 17012–190, Brazil
Tel. +55 14 3235 8247, E-Mail acm@usp.br

Author Index

Subject Index